Who Are We?

Who Are We?

ARE WE HUMANS GOD'S PERFECT CREATION?
If Not . . . Who Is? . . . and Who the Devil Are We?
Astounding Theory Takes Us Beyond Darwin!

Grace J. Nathan, M.S. and
John A. Friedman, Ph.D

To order additional copies of this book, contact:
Xlibris Corporation
1-888-795-4274
www.Xlibris.com
Orders@Xlibris.com
46242

CONTENTS

To the Memory of

Anthony Von Leewenhoek

and

Charles Darwin

AUTHORS' NOTE

The descriptions and theories in this work do not include opinions or beliefs about the creation of life—We believe these answers lie far beyond our human capacity to comprehend.

However, within the limitation of our human perception and understanding we accept that a prominent form of life exists on the earth and that the earth is a planet revolving around the sun and part of a solar system. A form of life exists here and we are a part of it. How it came into being is a matter for others to determine.

In our view, there is no conflict between beliefs about divine creation and the independent study of the characteristics and relationships of living organisms.

The record of life as it evolved and continues to evolve from the simplest of single celled organisms to the more complex is beyond doubt and should not conflict nor be confused with beliefs regarding how life was originally created.

The amazing commonality and relationship of life forms from single celled organisms to highly complex plants and animals is something at which to marvel.

In this work we hope to go beyond existing ideas about the nature of life. To this end we offer the theory we name *The On-Off Programming of Life on Earth.*

Acknowledgement

Our grateful thanks to Rollin Hunt for his invaluable editorial assistance.

SPECIAL ACKNOWLEDGEMENT

Our grateful thanks to Ms. Helena Curtis for her marvelous book, *The Marvelous Animals*. We found it to be an extraordinary resource for understanding and appreciating the importance of the marvelous world of the Protozoa.

CHAPTER I

Concepts Covered In This Work

This work sets forth a number of hypotheses, theories and assertions concerning the nature of life on earth. These concepts are briefly outlined below:

The earth was originally part of the sun consisting primarily of the sun's matter and energy. It revolves around the sun rotating on its axis facing the sun in an on-off digital cycle. This on-off cycle governs all life on earth. The authors believe it to be a universal law and have named it The On-Off Theory of Life on Earth.

Plants evolved as the primary form of life on earth appearing as a simple green plant-like form containing the chelating compound chlorophyll. They possess a unique affinity with the sun and the means to carry out photosynthesis.

The theory that animal life evolved from plant life is introduced here. Single-celled animal organisms originated and continue to evolve from single-celled plant organisms.

As Darwin aptly demonstrated, single –celled plant and animals continue to evolve into multi-cellular organisms becoming more complex and specialized through the process of natural and sexual selection and survival of the genetically fittest. Following in Darwin's footsteps is the Predator Principal introduced here. This principal establishes the nature of all animals and man as predators and recognizes the absolute and total power of the *Eat or be Eaten* predator instinct.

The extraordinary importance and consequences resulting from the incompleteness of human species programming are established.

A theory regarding the digital nature of time is introduced. It is proposed that time is a derivative of the on-off digital cycle of the sun as the planet rotates. Time is human perception based on the on-off presence of the sun.

The on-off programming has been inherited by all life forms. These are absolutely ruled by it.

The essence of psyche and the sense of one's self are established as the mystical and universal characteristic of all life.

The psychic nature of the on-off programming of life is also introduced.

The Theory

Theory proposes that life on earth functions in an on-off cycle based on the rotation of the earth and the "on-off" availability of the sun.

The theory emphasizes the vital importance of photosynthesis, the mysterious process that enables green plants to begin the manufacture of protoplasm and oxygen.

Plants are the perfect form of life on earth and are completely programmed for survival.

Animals evolved from plants.

Animals are an aberrant form of life and are imperfectly programmed for survival. They have lost their chlorophyll and the ability to manufacture their own food.

Animals and man are predators reduced to prey upon other forms of life to exist.

The existence of life on earth is totally dependent on plants and the On-Off Cycle of the sun.

The "on presence" of sunlight is psychically understood by all living organisms as the time to function and live.

The "off absence" of sunlight is psychically understood as a time of reduced function and the danger of death.

All living organisms are instinctively programmed to respond to the on-off presence of the sun.

Plants function in total harmony with the cycle of the sun. They produce their own food and oxygen. They are completely species programmed for survival. They are immortal.

Animals and man have lost this perfect affinity with the sun. They are incompletely species programmed for survival. They are unable to produce their own food and oxygen and must prey on other living organisms to exist. They suffer aging breakdown and death.

Authors' note: A spectacular example of how the on-off cycle regulates life is how human brain cells function in the on-off manner in thought. Also how dramatically this powerful on-off instinct has been applied to the current revolution in digital technology.

The authors are compiling a Psychic Language Dictionary which will reveal how language expresses our instinctive responses to this on-off programming of life.

————

What We Think We Know

The earth was originally a part of the sun. The elements and energies of the sun became those of the earth. The inorganic matter and energy on earth originated in the sun.

Although we will never fully understand the mysterious nature of life, we can assume that present forms of life and organic matter originated with green plant-like forms in the presence of sunlight and with the process of photosynthesis.

Can we also assume that this strange phenomenon we call life has a unique relationship with the sun?

This relationship is perfectly demonstrated by the plant in the process of photosynthesis. Photosynthesis began within chlorophyll-bearing tiny plant-like organisms which energized by sunlight, were able to create this enormously important process which sustains all life on earth.

————

Bits and Pieces

All of our earth consists of "bits and pieces" of the sun. What we refer to as matter and energy are "remnants" of the sun.

We believe the energy of the sun is everywhere and is in all things. It governs the earth totally. We are all—organic and inorganic, living and dead—remnants of the sun.

Life, living protoplasm, begins with the sun's materials re-activated by contact with the sun's light in the process of photosynthesis.

The tiny green plant with its chelating chlorophyll is able to perform the amazing process of simply shuffling atoms to produce simple sugars and free oxygen from carbon dioxide and water. This is the first step in the production of living matter and atmospheric oxygen.

Authors' note: We believe that life on earth is governed by, and continues to exist within, an instinctive response to the "on" and "off" presence of sunlight.

Of all forms of life, plants have the most complete and natural instinctive response to this powerful on-off cycle of the sun. They use it to create their own food and the oxygen needed to sustain themselves.

We believe that animals represent a "flawed" and unnatural form of life. They too are instinctively programmed to respond to this on-off law of sunlight but are unable to use it.

Chips Off the Old Sun's Block

As previously stated our planet and the other planets in the Solar System consist of bits and pieces of the sun.

Almost everything on the earth—matter, energy and life forms must therefore consist of, and be created from, the sun's "material". They were inherited from the sun.

Because of its absolute history with the sun, we believe that life has inherited a vital dependency on the sun. The continuing presence of the sun is essential for life to exist and survive.

Plants live in perfect harmony with the sun. They use its power to perform photosynthesis, the process that is vital to the creation and continued existence of all life.

The existence and survival of all life on earth depend exclusively on the unique relationship plants have with the on-off availability of sunlight as the earth performs its daily rotation.

In dramatic contrast to animals, plants might be considered nature's ideal life form and the earth's perfect resident.

Animals and man have also inherited a psychic awareness and relationship with the on-off cycle of the sun but they have lost their chlorophyll and their ability to harness the sun's energy. They have lost the ability to create their own food and have become the earth's predators.

These losses occurred during the evolutionary changes that produced the first protozoan single-celled plant-animals. Chlorophyll fell into disuse by some and the creative process of photosynthesis was discarded.

After all, why work at producing your food when there is an easy way of stealing it?

When borderline plant-animals became animals only, they abandoned their plant characteristics and became predators feeding on other organisms and each other.

All animals including man live as the earth's predators existing only by consuming other living organisms and each other.

―――――――

The Profound Significance of Photosynthesis

As previously discussed, early signs of life were thought to occur when tiny plant-like forms containing the chelating compound, chlorophyll, were able to capture and use the energy of the sun.

The tiny new "organisms", energized by sunlight, became diminutive engines capable of rearranging the molecules of carbon dioxide and water to create the simple. sugar, glucose and free oxygen.

They were then able to create more complex carbohydrates, fats and amino acids, the building blocks of protein, by adding and rearranging the atoms in sulfate and nitrate compounds available in the soil.

These newly formed compounds were then transformed into the living protoplasm of the plant.

The process of creating living protoplasm and releasing atmospheric oxygen by harnessing the energy of sunlight and using the basic materials of the earth begins with the amazing process called photosynthesis.

It should be emphasized that this vital process also produces the critically needed oxygen of the atmosphere making respiration and the oxidation of foods possible.

Photosynthesis is the most important chemical process on earth. It initiates and sustains all life. Without it life would cease to exist.

The On-Off Theory maintains that life on earth originates with plant photosynthesis which functions in an on-off recurring cycle in tandem with the earth's rotation and the on-off availability of sunlight.

This programming of photosynthesis in plant life instinctively follows the alternating on-off appearance and disappearance of sunlight in a natural harmony.

An imperfect form of this instinctive on-off programming was inherited by animals as they evolved from plants. Even humans are instinctively compelled to function in this "day-night", on-off sunlight, program inherited from plants and photosynthesis.

———

A Review of Plants, Sunlight and Photosynthesis

As previously indicated, the presence of sunlight is the power by which plants exist and function. Without plants all life including animals and man would cease to exist. The chemical process of photosynthesis is therefore the most important process on earth.

Again, it is in the presence of chlorophyll and the energy of the sun, that enables the plant to rearrange the atoms of water (H_2O) and carbon dioxide (CO_2) and create the simple sugar, glucose ($C_6H_{12}O_6$) and free oxygen (O_2).

$$CO_2 + H_2O + \text{sunlight \& chlorophyll} = C_6H_{12}O_6 + O_2.$$

The atoms in these simple sugars are then rearranged into more complex carbohydrates and fats and combined with minerals in the soil to produce amino acids, the building blocks of proteins.

It is in this way that the plant creates its own supply of food and its reserve of oxygen which it vitally needs in respiration.

It is the rotation of the earth and the on-off availability of sunlight that makes plant life and therefore all life possible.

The On-Off Law

Living things on earth function according to a fundamental "on" and "off" law based in the rotation of the earth and the alternating availability of sunlight.

This law was established by a primitive plant-like form able to harness the sun's energy and perform photosynthesis. This alternating on-off presence of the sun and photosynthesis became the established programming of all living organisms. It also was inherited by, and governs the species programming of all animal and human life.

Plants are the earth's creators—producing living matter from inert, inorganic compounds. This vital creative achievement begins with the process of photosynthesis and the harnessing off the sun's power.

Animals and man, on the other hand, are the earth's destroyers. They exist only as predators and owe their continued survival to the existence of plant life.

We will show in subsequent portions of this work that plants, as creators of life, are species programmed for immortality.

Animals and man, as destroyers of life, are species programmed for their own destruction and eventual death.

Is The Essence Of Life In Our Sun?

Photosynthesis and Immortality

Photosynthesis, the plant's ability to use the sun's energy to create life makes immortality possible.

As long as there is a sun and the earth with its atmosphere and minerals in the soil; as long as there is chlorophyll to carry out photosynthesis, plants can live indefinitely.

The species programming of plants is in perfect harmony with nature and is a totally effective method of survival. Plant life programming is immortal.

When plants lost the special affinity with the sun and ability to perform photosynthesis and became animals, they became a less perfect species.

They lost their ability to create the food materials needed to survive. So to live, they became predators hunting and consuming other living organisms and even each other.

The predator species programming of animals and humans is at odds with nature. It is an incomplete, complex and defective method of living. The programming of animals and humans produces deterioration with age and death.

Good and Bad Species Programming for Survival

We believe that life exists as a response to the powerful influence and impact of the sun.

We believe that the degree of effectiveness in the programming of a life form depends on its degree of its affinity with the sun.

Plant Life

Plants are the most completely species programmed for survival.

They may be considered the "perfect" form of life. They function in perfect harmony with the on-off cycle of sunlight. They are the original farmers—utilizing the energy of the sun, the minerals in the soil and water and carbon dioxide in the atmosphere to create their food and the oxygen needed to sustain life.

They possess the greatest affinity with the sun and are completely species programmed for survival.

Animal Life

Animals are inadequately species programmed for survival.

The creative instinct directing the production of food and oxygen has been replaced by the destroying predator instinct.

The psychic awareness of the sun has been replaced with the awareness of the animal's own self and the self of the predator. The predator instinct—Eat of be Eaten—has emerged as a new way to live.

Human Life

We humans have the most unorganized program for survival of all species.

We are born approximately nine months too soon. We are born before the programming of our species can be completed. We are born with our instincts intact but with no effective programming to satisfy them.

Consider how our infants arrive at birth. They are in a virtually helpless state. Their brain and physiological systems are undeveloped and incompletely organized. The infant requires years of physical and psychological care to survive and reach maturity.

The human brain and physiological systems are undeveloped at birth and must develop and become organized after birth.

This precipitous-early birth was necessary to allow the passage of the human head (which will greatly develop in size) to pass through the small birth canal.

The head (with its superior brain) must be released prematurely from the womb in order for both the baby and mother to survive.

We humans are the most incompletely programmed of all species.

However, to survive our early birth and ineffective species programming, we evolved the most highly developed brain and psychic awareness of all life forms.

We also, it should be noted, like that of all animals, are totally programmed to live as predators. Our psyche signals when conditions are right for getting our "prey" and signals when we are threatened by a "predator".

The predator instinct, eat or be eaten, rules animal and human life.

The On-Off Life Cycle in Plants and Animals

The plant world exists in a perfect harmony with the sun.

Plants function in tandem with the on-off cycle of the sun's presence. They create the compounds to be used to create their own protoplasm and produce a reserve of oxygen to be used in respiration.

This is performed during the "on sunlight" period. During the "off sunlight" period, photosynthesis ceases and the plant reduces functions.

Animals (including man), on the other hand, can create no food nor oxygen. They have become an incomplete form of life that has lost its ability to produce food and must live as predators feeding on other life forms and breathing the oxygen generated by plants.

The world of animals and humans is ruled by the predator instinct. The predator instinct, Eat or Be Eaten, is the absolute force directing all animal and human activity.

The universal on-off cycle of life, inherited by animals and humans from their plant ancestors, became adapted to serve the predator instinct. Photosynthesis, the creative "on" phase of the cycle, was replaced by the destructive "on" instinct to hunt and devour other living organisms.

The original "on" signal to create life became the "on" signal to capture and destroy life.

―――――――――

A Defining Characteristic of Life

The characteristic common to all life and that which separates the living from the non-living is its *affinity with the sun*.

Plants

Plants live and function in perfect harmony with the sun. They possess the greatest degree of awareness of, and affinity with, the sun.

They are perfectly adapted to the earth's rotation and the on-off presence of the sun.

They are able to sustain themselves by producing their own food and oxygen.

Their species programming is completely adapted for survival.

They are programmed to live forever.

They have a high degree of awareness of the sun and seem to have less need for an awareness of the self.

Animals

Animals live and function in less harmony with the sun. They possess less awareness of, and little affinity with, the sun.

They are unable to produce their own food and oxygen and must prey upon other forms of life to survive.

Their species programming for survival is at odds with nature. It is flawed causing aging and eventual death. They possess an increased awareness of themselves and their predators (an adaptation needed to hunt prey and to protect themselves).

Humans

We humans live and function in least harmony with the sun. We possess very little awareness of, and affinity with, the sun.

We are born too soon. We are born helpless and unable to survive without extensive physiological and psychic support during our infancy and childhood.*

We are unable to produce our own food and must prey upon other forms of life to sustain ourselves. Our species programming for survival is woefully incomplete. Like animals, our species programming is flawed, causing aging and death.

---

* Reference *Are We Humans Born Too Soon?*

We possess a high degree of self-awareness and the awareness of others. We also possess a great capacity for psychic creativity. Both are highly developed adaptations needed for our survival.*

The On-Off Programming of Awareness

Although all forms of life on earth function in a tandem program with the daily on-off presence of the sun, it is the plant that functions in genuine affinity with, and awareness of, the sun.

It is the plant that is capable of capturing the essence of sunlight and putting it to use in the process of photosynthesis. It is the plant that creates, maintains and holds the secret of life.

Life's Psychic Awareness

We believe that the unique psychic awareness present in all life is inherited from, and originated in, the plant's psychic awareness of the sun.

All psychic awareness therefore appears to be a replication of the plant's awareness of the sun and the sun's on-off programming of availability.

As noted, the plant uses its awareness of the on-off programming of the sun to create life.

We believe the psychic awareness of animals and humans was inherited from plants and was applied in a destructive way. The photosynthetic instinct was replaced by the predator instinct.

The on-off programming of the photosynthetic way of life. was replaced by the "Eat or Be Eaten" predator way of living.

The Psychic Meaning of "On-Off"

The "on-off" presence of the sun has great psychic significance.

To the plant, the "on" phase, or appearance of the sun, signals the ideal condition needed to live. The "off" phase, or absence of sunlight, signals danger to life.

On represents Life. Off represents Death.

In the process of photosynthesis, the plant precisely follows this "On Life-Off Live" cycle as it grows and creates its own protoplasm.

The Awareness of Self

Another major transformation took place as the animal evolved from the plant. The plant's acute "on-off' awareness of the sun became adapted to serve the predator instinct.

The "on" phase continued to represent good conditions but became a good time for hunting prey. The "off" phase continued to represent reduced activity in the form of sleep.

On represents Life. Off represents Sleep.

An increase in the awareness of the individual self and awareness of the predator also replaced the awareness of the sun.

Animals had lost their connection with the sun and ability to create their food. They had become greatly aware of, and able to identify, their prey and their predator.

Animals and man acquired a heightened psychic awareness of self and the self of others in order to insure their survival as predators.

We believe that man, because he is born too soon and is species incomplete, has had to develop this greater degree of psychic awareness and psychic creativity in order to survive.

Plants are highly psychically aware of the sun. Animals are aware of the self and are highly aware of their own specific predator.

Humans are highly psychically aware of individual self and the self of others. They also have the unique ability to create imaginary perceptions of reality in the form of illusion, fantasy and delusion.

Note that an explanation of how this unique and high degree of psychic perception developed in humans is more fully discussed in the authors' work, *Are We Humans Born Too Soon?*

Species Completeness For Survival

Plants are the most completely species programmed for survival. Their programming functions in perfect harmony with the sun. They create their own food materials and are self-sustaining. They are immortal. They are species complete.

Animals are abnormally species programmed for survival. Their programming imitates the plant's relationship with the sun but they can only sustain themselves by destroying life. They are programmed to experience physical and mental breakdown and eventual death. We regard them as species complete but aberrantly programmed.

We humans appear to be both the most aberrant and incomplete of all living species. We are the only living species that is born too soon with any chance of survival. We are born before our brain and physiological systems can become organized enough to function.

For a more complete explanation of the species programming of humans, please refer to the authors' work, *Are We Humans Born Too Soon?*

_____

Instinctual On-Off Functions

All living organisms, since they are descendants of photosynthesis, instinctively function according to the on-off presence of sunlight as the earth rotates.

In plant life all physiological processes and instinctive responses are in perfect harmony with the on-off availability of the sun's light.

We will show that the physiological processes and instinctive responses of animals and humans were inherited from plants and are also influenced by this natural law.

The "On Sunlight" phase instinctively signals the "Life is on" message.

In humans for example, the "on sunlight" phase is experienced psychically in degrees of pleasurable emotions.

The "Off Sunlight" phase instinctively signals the "Life is turned off" message.

In humans, darkness represents danger of death and is experienced psychically in degrees of painful and fearful emotions.

_____

Psychic Awareness and Photosynthesis

We have established that psychic awareness of the individual self and the self of others are defining characteristics of life. Awareness of self ceases when life ceases. We humans say that our soul has departed.

We believe this psychic awareness originated with the plant's awareness of the presence of the sun and the phenomenon of photosynthesis. The plant's awareness allows it to perform photosynthesis in perfect on-off harmony with the sun—functioning with sunlight and ceasing or reducing function in darkness.

We believe that psychic perception, the force that characterizes and is unique to all life, began in a tiny plant-like organism with the special chelating ability to bond with the power of sunlight making the process of photosynthesis possible.

Plant life functions in perfect harmony with the on-off cycle of the sun.

Animals and humans, the imperfect offspring of plants, do not function in harmony with the sun's on-off cycle. They are unable to carry out photosynthesis and have distorted the on-off cycle to accommodate the predator-prey instinct.

All life on earth depends upon the plant's affinity with the sun and its ability to perform photosynthesis. Without this all life would cease to exist.

Plants are the creators of life functioning in harmony with nature.

Animals and man are destroyers of life destined to live in conflict with nature.

Life's Unique Psychic Awareness

The single most characteristic common to all life is psychic awareness. All living organisms appear to possess an organized "awareness" of themselves. The organism is dead if there is no awareness.

We believe this psychic awareness was first experienced as an awareness of and special affinity with the sun. A single celled plant-like "organism" containing chlorophyll reacted to the presence of sunlight.

Self-awareness appeared in the presence of sunlight and disappeared with the absence of sunlight.

We believe the plant's psychic awareness of the sun is the vital life force that separates the living from the non-living. Life began with this special affinity with the sun and the ability to "bond with" the power of sunlight in the process we call photosynthesis.

We believe that all life exists because of the plant's psychic awareness and affinity with the sun.

Can we also assume that the psychic awareness common to animal and human life is inherited from the plant's psychic affinity with the sun?

Can we therefore conclude that the life force that exists in animals and humans is also inherited from the life-force of plants and their ability to capture the power of the sun?

If psychic awareness is inherited from plants and the process of photosynthesis, it too must function in the on-off cycle with the earth's rotation.

We call this awareness of the sun The On-Off Programming of Life on Earth.

Plants exist in perfect harmony with the On-Off Programming of Life.

Animals and man are unable to use the sun's energy and are therefore unable to live in harmony with the sun. They have lost the ability to produce food and oxygen and must live abnormally as destructive predators.

The continued existence of life on earth depends upon the presence of sunlight, green plants and photosynthesis. Without sunlight, plants and photosynthesis, all life on earth would cease to exist.

Plants are the earth's life-givers. They produce food and oxygen and the "food" and oxygen needed by all other life forms.

Animals and man are the earth's predators unable to produce their own food and oxygen and compelled to exist by feeding on other living organisms and each other.

CHAPTER II

The Plant-Animals

*The species programming of animals recapitulates
the phylogeny of plant photosynthesis.*

The Incredible Protozoa

"Two hundred years ago, almost every gentleman had his own microscope which was as much a mark of his cultivation and interest in the world around him as the richly bound volumes that lined the walls of his library or the handsomely mounted globe that stood by the window that lead into the garden.

For example in one eighteenth century melodrama, the young heroine regretfully declines the pleadings of an importunate suitor who wishes to carry her off with a cry of "What! and leave my microscope!"

So protozoology not always demanding of so noble a sacrifice remains first and foremost an amazing and incredibly important sturdy.

There are important reasons for studying these amazing creatures. At about the time that Koch and Pasteur established that bacteria cause disease, the protozoan became associated with amoebic dysentery and malaria, and since that time the protozoa have been investigated for medical reasons. Some of the diseases for which these parasites are responsible are serious medical problems, but actually considering the enormous numbers and ubiquity, these one-celled plants and animals are surprisingly harmless. In fact, they may be considered mush more as friends and allies.

Protozoa are a basic part of the long food chain to which we contribute nothing and on which our survival depends, as well as that of the rest of the animal world. The protozoa are also a major asset in the struggle against the pollution of our natural waters. A single paramecium can devour five thousand bacteria in a day.

They are Everywhere

Protozoa are found in the fresh waters of mountain streams, in brackish swamps, in rain barrels and in sewers. Many live in the soil about five or six inches below the ground, their entire universe consisting of the thin film of water between particles of earth. The most common and numerous inhabit the ocean waters forming much of the plankton, the rich meadow of the sea, on which the entire chain of marine life depends. Ocean travelers can recall nights when the ship's wake shone for miles with incandescent light and seaside dwellers remember evenings when a wading child traced an enchanted moon path trough the water—the result of the flashing lights of millions of protozoa.

Many common species are found everywhere in the world. A pond in California may yield the same protozoa population as a similar pond in England or in China. The organisms found by Leeuwenhoek in the small inland lake near Delft three hundred years ago are likely to be there today and in lakes on every continent. Frogs all over the world carry similar protozoan parasites in their intestines and sea urchins from every shore of every continent are hosts to same species.

One reason for the ubiquity and hardiness of many protozoa is their ability to form protective cysts. In this state, they may be carried by the winds for miles, across oceans and continents, or they may remain dormant in the same spot, clinging to a germ of life and waiting to be awakened.

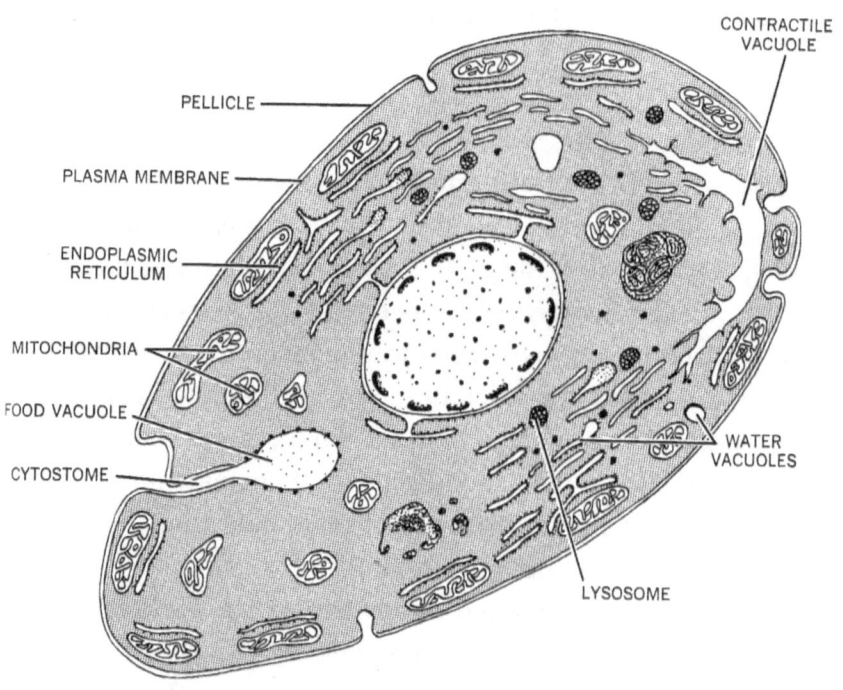

Figure 1 Diagram of the "digestive system" of the ciliate Tetrahymena (a plant).

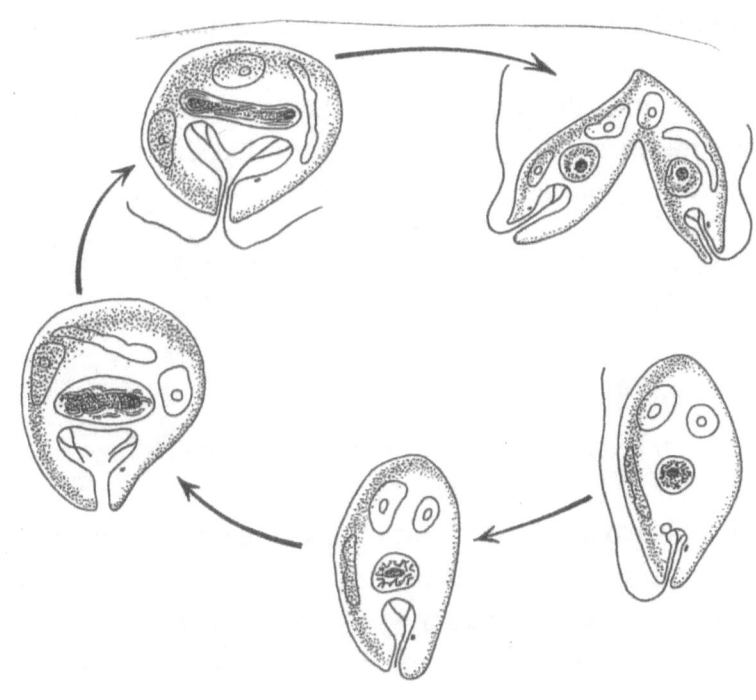

Figure 2 Euglena and other flagellates divide longitudinally, each new cell forming a mirror image of the other.

There are more than thirty thousand species of protozoa living today everywhere among us. They represent billions of individuals, far more than all the other animals in the world combined.

In one small pond the various creatures inhabit their own microcosms, the green chlorophyll-bearing plantlike forms on the surface, amoebas crawling on the bottom, some species clinging to the sides, others swarming around bacteria or fastened to bits of plants or animals. One type of protozoa lives on mountain peaks forming a red film on the snow another in the near-boiling waters of Hot Springs, another in the ice floes of the Artic. Even the desert has its protozoa spending their existence in dried shells and emerging once a year for an hour or so with a rainfall.

Which are Plants? Which are Animals?

All of them are one-celled. Some of them are clearly animals; that is, they do not possess chlorophyll or similar pigment and are not therefore capable of photosynthesis. Others do possess chlorophyll and do carry out photosynthesis. Most wiggle around and *are so closely related to each other that it becomes impossible to separate the two groups.*

The reason for this dilemma can be more clearly understood if we recall that the classification of living things reflects their evolutionary history. There were probably so many changes and adaptations that there is no common ancestor available to us. For example, even certain algae, lost their chlorophyll and joined the animal world quite late in their evolutionary history and it is probable that certain protozoa may have acquired photosynthetic capacities becoming plant-like algae as well.

The protozoans, although only one-celled, contain extremely complex organelles. These are being studied today and indicate that whether they are in the single-cell of plants and animals or in the cell of our own body, they are alike.

Each protozoan cell contains the features common to all cells; mitochondria, nuclei, chromosomes, membranes, lysosomes,—all are constructed along the same general design and repeated over and over throughout all forms of life.

Protozoans excite our curiosity because they offer a glimpse of the world of billions of years ago. Their ancestors were early experimenters in living, and with them the most fundamental problems of survival had to be worked out; namely,

How to get food,

How to assimilate it;

How to store it;

How to eat and ingest another animal; even its twin, without accidentally digesting itself;

How to repair and coordinate all of these necessary functions;

How to reproduce its own kind, providing the correct balance between genetic continuity and diversity.

A visible index of the extent of this experimentation is apparent in the variety of the protozoa now in existence and the extraordinary range of environments in which they live.

The protozoa tried everything and in so doing, adapted themselves to the most improbable ecological niches. Only the bacteria live in more varied and improbable habitats.

The proliferation of the one-celled organisms and the long span of time in which they have evolved—and are evolving even today— provided an unequaled opportunity for the great variation and for the forces of natural selection to pick and choose, editing out mistakes and reinforcing successes. Many of these records of experimentation have been lost forever.

Living protozoa are not in any way ancient however. They are as modern as you and I. Still, there are clear indications that some organisms have remained unchanged through the millenniums—since long before the thunder lizards appeared on the planet.

Survivors today provide fundamental answers to questions regarding the evolution of plants and animals.

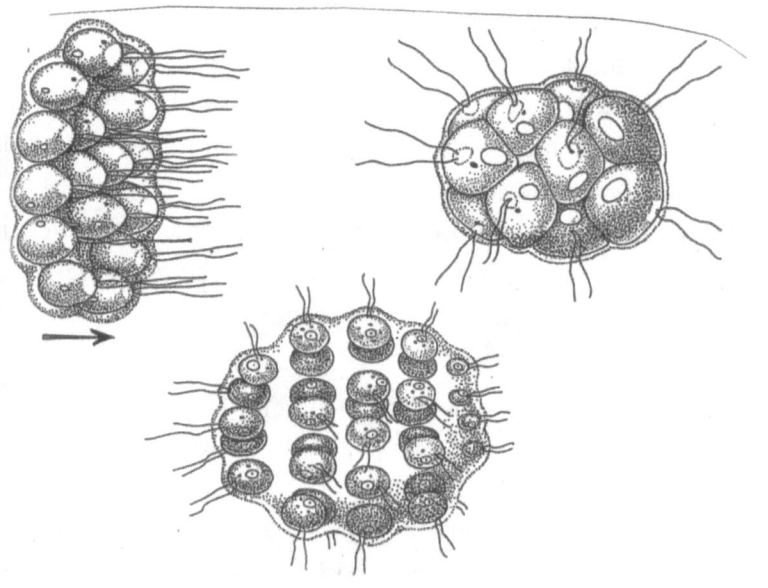

Figure 3 At some unrecorded point in time, perhaps very early in the history of protozoa, certain of the one-celled animals began to form societies of cells—communities in which different cells had different functions—and from these societies the metazoa, the many celled animals, are believed to have evolved.

Feeding Habits of the Protozoa

Protozoa get their food in three ways:

By photosynthesis involving harnessing the energy of sunlight in the presence of chlorophyll.

By the saprozoic method of absorbing substances dissolved in the water environment through the cell membrane. This method is much more complex than it appears and is basically the way the cells in our body are nourished.

By phagocytosis consisting of simply devouring other organisms.

Many protozoa feed in all three ways—sometimes through photosynthesis like plants and sometimes by phagocytosis like animals capturing and engulfing another protozoa.

Most of the saprozoytes are parasites. For example, Opalina, an attractive little animal, lives entirely in the digestive tract of frogs. Opalina's ancestors probably spent millions of years developing mouths or cytostomes, before they moved into the frog but the many species of this ubiquitous but harmless parasite spent additional millenniums losing all traces of the mouth and now live an exclusive saprozoic existence.

The Flagellates—Are They Plants or Animals?

The phytoflagellates versus the zooflagellates. Many are both.

In a single drop of water from the surface of a pond and a good microscope, one can get a glimpse of what the world might have been like a billion and a half years ago.

There will probably be several patches of algae darting through the water so swiftly you will barely see them as tiny swimming specks. These animals or plants are small flagellates.

No one knows since no one was around some 1,498,000,000 years ago, but some of these modern creatures are thought to be the ancestors of the *first animal forms on earth.*

The first true cells to evolve were probably much like Chlamydomonas, a common but highly complex and organized little green flagellate. Chlamydomonas like other volvocidae have a rigid cellulose wall like the cells of a flower and is the most plant-like of the flagellates.

These little creatures are clearly plants and it is generally agreed that the multi-celled plants evolved from phytoflagellates like these which settled down, lost the flagella, and took up a social form of photosynthetic existence.

Originally botanists classified them as an alga, but these flagellates are so clearly related to animals as well that the zoologists firmly declined to yield them over.

The Chrysomonads are also small chlorophyll containing flagellates that multiply so rapidly they often cover the surface of a pond. They, however, lack the stiff cell wall and may often assume amoeboid shapes putting forth pseudopods which they use to move and feed.

Many of them are phagocytic as well as photosynthesizers assuming both animal and plant identities. It is not unusual for them to lose their flagella and "round up" into a sphere stage that is indistinguishable from algal plants. Others are known to lose both their flagella and their chlorophyll and become completely amoeboid and clearly animal-like.

The Chyrsomonads form an important part of the micro plankton of both salt and fresh water; they are often found together in large clearly visible colonies moving as a mass on the surface of the ocean.

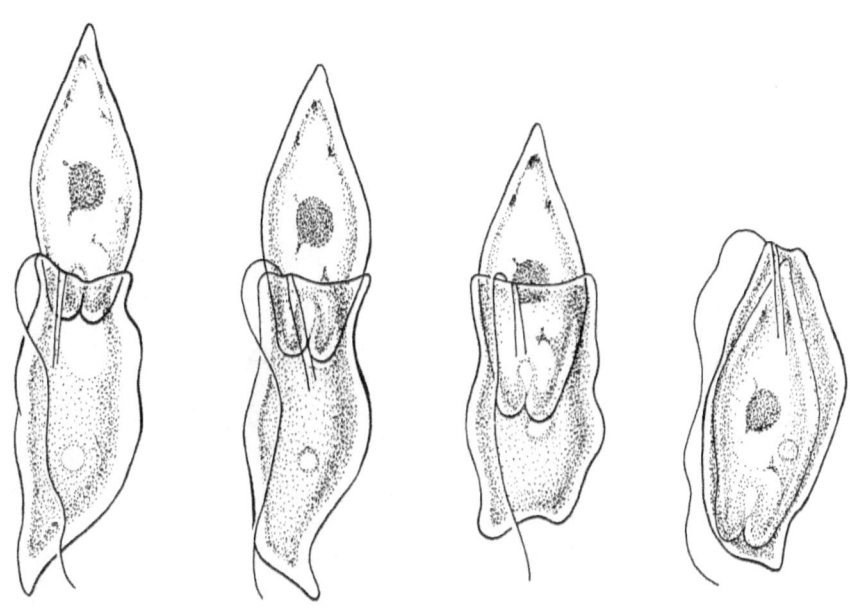

Figure 4 Evolved from a common ancestor, look-alikes paranema (an animal) devours euglena (a plant)! With cousins like this, who needs enemies?

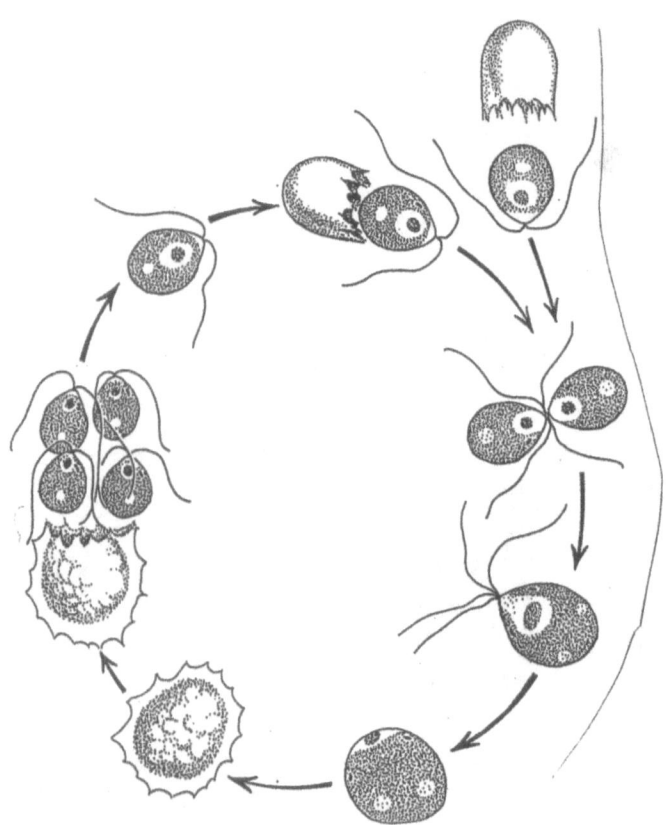

Figure 5 The beginnings of sexuality. Chlamydomonas, a small plant flagellate, which often reproduces asexually by simple cell division, also demonstrates early sexual evolution. Unique among other protozoa and other multi-celled animals, Chlamadymonas is naturally haploid, that is, its chromosomes do not occur in pairs. Most often it simply undergoes binary fission, forming two new mirror-like images. Alternatively one may encounter another which, although it looks identical, is apparently different, as revealed by their instant attraction to one another.

———

The Dinoflagellates

Among the chief contributors to the ocean plankton are the Dinoflagellates, some of which are naked and some have a variety of cellular wall shapes. Their chlorophyll is often masked by other red and brown pigments and when they form in large numbers they color the sea. The Red Sea gets its name from these nomadic creatures.

The Dinoflagellates may on occasion become algae-like and plant-like losing their flagella or they may lose their chloroplasts and become amoeboid and animal-like.

———

Euglena and Other Euglenics

The euglenoids include species that are permanently plants that are photosynthetic. They also include forms that feed by phagocytosis and so are clearly animals yet the unity of the order is undeniable.

Typically the euglenoid has an elongated body with a reservoir pocket at the anterior end into which the contractile vacuole opens. A flagellum protrudes from the reservoir *with an eyespot at its base in photosynthetic forms.*

Most important of all is that the Eugenloids store their food *not as starch like plants nor as sugar as animals do, but as a starch-like substance called paramylum* which is unique to this order of organisms.

Euglena has a host of cousins. One is phacus which has a single flagellum, a flat disc-like green body that spins through the water like an animal.

Another of Euglena's close cousins is Paranema. Paranema is almost identical in structure to Euglena but Paranema contains *no chlorophyll and is definitely an animal-like phagocyte* feeding on all sorts of organisms including its cousin, Euglena, one of its favorite foods.

———

Protozoan Reproduction

We humans produce a modest number of offspring compared to almost all other animals but we tend to survive as individuals.

Any single paramecium, on the other hand, has little reason to expect a long life. However, each single one can replicate itself so rapidly as the touching off of a chain reaction—two, four, eight, sixteen, thirty-two, sixty-four, one hundred twenty-eight—that soon reaches an astonishing total. In fact, one patient protozoologist once calculated that if all the progeny of any single paramecium survived (assuming a division rate of once a day), in 113 days there would be a mass of Paramecia equal to the volume of the earth. Some species, in fact, can divide as often as five times a day, which would greatly hasten the process. It is this high rate of reproduction that permits the survival of the one-celled plants and animals and in all probability, the development of the many species and varieties of protozoa."*

———

---

* This section The Incredible Protozoa is reproduced from *The Marvelous Animals An Introduction to the Protozoa* by Helena Curtis published by The American Museum of Natural History, Garden City, NY, 1968.

The Appearance of Animals

As previously suggested, animal life evolved (and continues to evolve) from aggressive, single-celled plant-like organisms called Flagellates.

These Flagellates lacked the green collating chlorophyll material of their plant cousins. They were therefore unable to create their own food and oxygen needed to live. They had abandoned their relationship with the sun and their perfect photosynthetic method for survival.

They, the new animals, could no longer sustain themselves. They had lost the method of living in a perfect harmony with the sun and the earth's elements.

The new single-celled animals would change the world. They became predators who could only survive by feeding on their relatives and each other.

The new world, formerly inhabited only by its green plant residents living in harmony with the sun, became a dangerous and destructive world of predators.

———

Animals From Plants

As previously stated in the section describing the Protozoa, we believe animals and humans evolved from ancestral single-celled plants that had lost their chlorophyll and their ability to carry on photosynthesis. They could no longer sustain themselves and had become predators able only to exist by preying on other forms of life.

Having evolved from plants, animals inherited many characteristics of plant programming. The awake/sleep cycle of animals, for example, is doubtless an example of legacy from the ancestral cycle of photosynthesis.

With some nocturnal predatory adaptations, animals and humans generally follow the photosynthetic program, functioning during the sunlight cycle and resting during the dark phase of the cycle.

The on/off cycle of sunlight moving with the rotation of the earth is life's basis of existence. It is instinctively "understood" by all living organisms. "On sunlight" represents active, positive conditions supporting life. "Off sunlight" represents negative conditions and the danger of death.

In the plant world of photosynthesis, the on-off cycle represents an opportunity to create life in harmony with the sun and the earth.

In the animal world, on-off cycle has become distorted to represent opportunities for capturing and consuming other organisms and the danger of being captured and eaten by a predator.

The destructive predator instinct, eat or be eaten, had replaced the creative photosynthetic instinct.

———

Nature's True Champions

Nature's perfect creation is certainly the plant.

Plants and a process called photosynthesis provide the food and oxygen needed to sustain all animal and human life. Plants make it possible for the predators to survive.

———

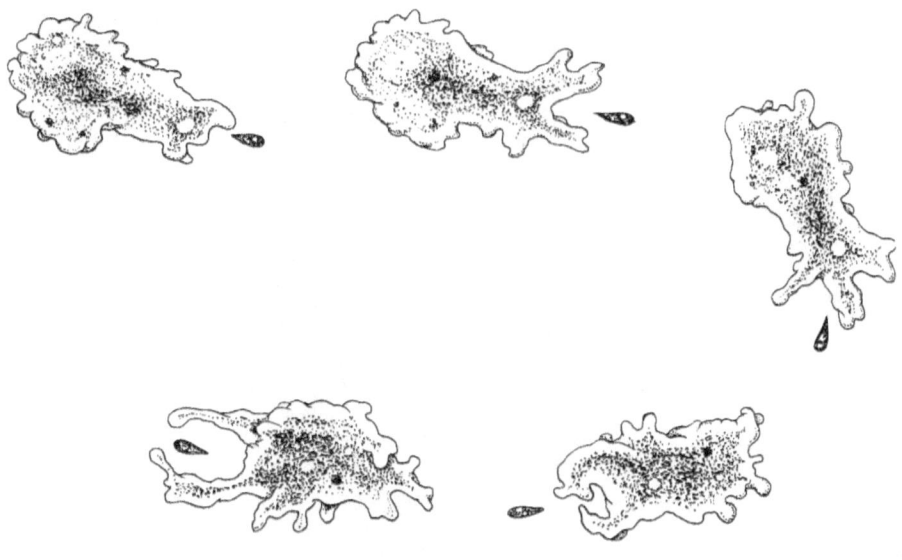

Figure 6 Amoeba (an animal) in pursuit of a small plant flagellate.

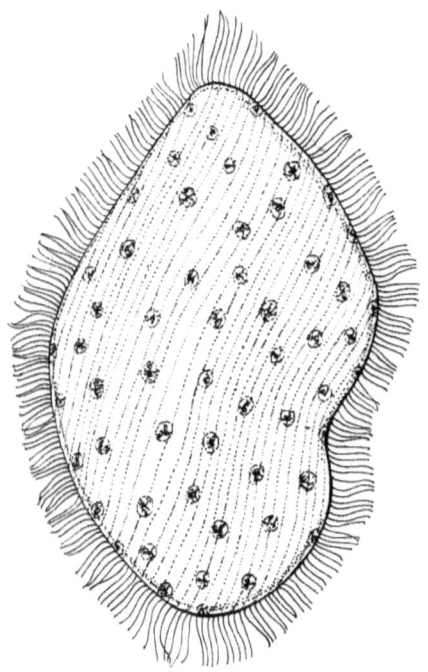

Figure 7 Opalina, which evolved into an animal protozoan once had a
mouth and now simply absorbs its food from the
digestive tract of the common pond frog.

CHAPTER III

Descent of the Predators—"Eat Or Be Eaten"

Photosynthesis—The Natural Way Of Living

This perfect way of life functions in harmony with the planet and the sun. It uses the sun's energy and the earth's inorganic elements to produce food, release free oxygen and support life.

Predation—The Unnatural Way Of Living

Although this aberrant and destructive way of life mimics the harmony of the plant and the sun, it can create nothing and must prey on other living organisms to exist.

———

The Evolution of The Predatory System

When certain single-celled plants ceased to be plants and became animals, drastic evolutionary changes occurred as they learned to feed on other living organisms and become predators.

This transformation from the plant's photosynthesis to the animal's "kill and eat" way of existence required billions of years of evolutionary experimentation and the force of natural selection to edit out mistakes and reinforce successes.

As previously noted, the ancestors of animals were early experimenters in living. The most fundamental problems of survival had to be worked out; namely,

How to get food,

How to assimilate it;

How to store it;

How to eat and ingest another animal; even its twin, without accidentally digesting itself;

How to repair and coordinate all of the necessary functions;

How to reproduce its own kind, providing the right balance between genetic continuity and diversity."*

In addition, effective systems for all of the above functions had to be successfully developed and learned.

Compared to the plant's life, the animal's predator method is deviant, extremely inefficient, complicated beyond belief and destined to break down.

Remember that plants did not need billions of years to learn how to sustain themselves. Their many evolutionary changes were primarily adaptations to changes in the environment.

---

*   From *The Marvelous Animals*

In contrast, the plant's totally efficient method of living creates rather than destroys life. In concert with the earth and the sun, it uses the power of sunlight and the inorganic compounds available in the atmosphere and in the soil to create its life.

The plant is able by using the energy of sunlight through the process of photosynthesis to miraculously rearrange the atoms in the molecules of carbon dioxide and water to form the molecules of simple sugars and free oxygen.

The plant is then able to build these simple sugars with other compounds in the soil into amino acids and the other essentials of living protoplasm.

Unlike animals, plants had no need to learn how to hunt and digest other living organisms. They are able to sustain themselves and grow by producing what they need in a natural and non-destructive way.

Two Methods of Living
Photosynthesis and Predation

The Marvel of Photosynthesis

"The following changes appear to take place in green plants under the influence of sunlight:

6 molecules of carbon dioxide + 6 molecules of water   form 1 molecule of glucose   + 6 molecules of oxygen

$$6\ CO_2 + 6\ H_2O = 6\ C_6H_{12}O_6 + 6\ O_2$$

Although six molecules each of carbon dioxide and water are enough to form each molecule of glucose, the number of oxygen molecules is more than enough for the glucose molecule. This excess of oxygen is liberated from the plant into the atmosphere.

The energy of sunlight is not lost but is stored in the glucose molecule as chemical energy and as we shall soon see, be recovered and used. It is released when glucose or related substances which the plant may produce from glucose, are broken down again into carbon dioxide and water in the process called respiration.

The plant may further convert glucose into another more complex sugar, $C_{12}H_{22}O_{11}$, or into the closely related carbohydrate, cellulose, used to thicken the cellular wall. Part of the glucose may be transformed into starch and held as reserve food in the cell. Fatty substances needed by the protoplasm are produced from the sugar. Proteins are also formed by uniting part of the sugar with nitrogen, sulfur and phosphorus in inorganic salts absorbed from the water of the river or lake.

In the formation of the fats and protein there is merely a rearrangement of the molecules of the sugar and there is little or no loss of energy.

***

Respiration

This process goes on continuously in all living plants and animals.

During respiration the molecules of glucose and other related substances are broken down into carbon dioxide and water releasing the stored chemical energy into forms that can be used by the living protoplasm.

Plants respire using the oxygen they have already produced. Animals respire using the excess oxygen produced by plants and released into the air.

Thus it is that all of the energy needed by plants and animals to carry on their activities is derived from the sun. This vital energy is stored in the food manufactured by photosynthesis and liberated by respiration.

If we consider glucose the substance manufactured by photosynthesis and oxidized by respiration, the equation for respiration becomes just the reverse of photosynthesis!

$$C_6H_{12}O_6 + 6\ O_2 = 6\ CO_2 + 6\ H_2O.*$$

———————

The Deadly Predator System

Predation is the predator system by which all animals and humans live. Predation is the antithesis of the photosynthetic system of living described above.

The predatory method of living requires the creation of a great number of unnatural physiological and psychic systems. These evolved as they became needed to solve the many problems involved in living in an entirely new way.

These problems were first solved by the single-celled protozoa animals. Their endless trial and error experimentation resulted in the many basic functions used in the complex physiological and psychic systems of today's multi-cellular animals and man

The Predatory System of animals and man is an ineffective and unnatural way to exist. It is an entirely dependent and destructive method of living destined to break down and die.

The Photosynthetic System of plants is in direct contrast to the predator. It is a perfect way of life—working in harmony with the sun

———————

*   Text from *Introductory Biology* by Andrew Stauffer
    Copyright 1949 by D. Van Nostrand Company, Inc. Canada, Ltd.

and earth and creating the food and energy needed to support continuing life. The system is efficient, natural and immortal.

––––––––

The Makeshift Predator

As mentioned previously, the single-celled animals were the first to be challenged to survive without photosynthesis. In order to survive, they had to develop working methods vitally needed to perform all sorts of new unnatural functions.

Through trial and error experimentation, the first predators had to devise both physical and psychic methods to capture their prey, ingest it, store it, digest it, assimilate its nutrients, eliminate the waste and learn how to reproduce themselves.

Consider the complications and bizarre challenges the multi-cellular animals had to face as they evolved and struggled to survive. Billions of experiments had to occur during each stage of higher, more complex development.

Consider that the predator's lifestyle is in direct, destructive opposition to the creative force of nature as exemplified by photosynthesis.

The physiological and psychic predatory systems that were developed required endless changes and modifications to accommodate the more advanced animals and man.

The systems in the higher animals and humans became so complex in their construction and so sadly inefficient as to be preposterous. One recalls the complex Tinker Toys of childhood.

Remember that the plant's method of living, photosynthesis, is self-sustaining, self-perpetuating and requires no maintenance.

Consider the enormous complexities of the predation systems needed by higher animals and man to maintain their predatory "eat or be eaten" lifestyle.

Is it any wonder that these incredibly multifaceted systems eventually break down in death?

Following is a review of some of the elements needed by these complex systems to serve highly developed predator lifestyles:

Skeletal System needs skull, teeth, spine, ribs, pelvis, inner ear, sinuses, arm, leg, feet and joint bones.

Muscular System needs hundreds of muscles in the limbs, body, eyes, head, neck, shoulders, torso.

Digestive System needs salivary glands, pharynx, esophagus, stomach, liver, gall bladder, pancreas, intestines, rectum, and anus.

Circulatory System needs arteries, veins, capillaries, heart, and blood cells.

Respiratory System needs nasal and aural cavities, larynx, bronchi, lungs, alveoli, and capillaries.

Urinary System needs adrenal glands, kidneys, ureter, bladder, and urethra.

Endocrine System needs hypothalamus, pineal, pituitary, thyroid, parathyroid, thymus, adrenal glands, ovaries, and testicles.

Integumentary System needs epidermis, dermis, subcutaneous and muscle tissue sweat glands, hair follicles.

Nervous System needs cerebrum, cerebellum, medulla oblongata, spinal cord. and millions of neutrons.

Reproductive System needs uterus, ovaries, vagina, and labia. penis, scrotum, prostate gland, seminal vessel.

Animal Species Programming for Survival

The species programming for animal and human survival evolved as an aberrant form of the plant's programming of photosynthesis.

However, the animal survival program lacks the crucial ability to create food and oxygen. It was also effective for only a limited period of time, eventually breaking down and resulting in the death of the animal.

The programming of survival of all animals and humans is driven by the predator instinct. For the species to survive, its program must be highly efficient at hunting and getting food and also at protecting oneself from being captured and eaten.

The evolution of the most successful predator program for each species depended on getting the suitable predator, and prey in the territory to which both were well adapted.

The evolution and natural selection of a survival program also needed efficient psychic programming to enable the predator and prey to recognize each other.

Natural selection and survival also required that predator and prey be matched in their ability to survive both each other and the physical demands of their environment.

The Predator Instinct—"Eat Or Be Eaten"

The Predator Instinct is the primary instinct governing all animals and man.

The instinct to create life through the process of photosynthesis was replaced by the predator instinct.

The predator instinct is programmed to destroy life. The predator that consumes its prey is also the prey to another predator. It became necessary for the animal to not only successfully hunt and capture its prey but to successfully avoid being hunted and captured by its predator. Eat or Be Eaten became the basic law of survival in the animal and human world.

In contrast, the species programming of plants creates and supports the continuum of life and is therefore immortal.

Note that the species programming of the animal and human species destroys the life force continuum, and is destined to break down in death.*

---

*   *Authors' note:* We perceive this to be true but don't quite know why. However, the predator instinct programming is so abnormal, so badly constructed and so complicated in functioning, it is destined to break down no matter how well maintained. In contrast, the species programming of plants is simple and self-generating, requiring no maintenance.

CHAPTER IV

The On-Off Programming of Psychic Awareness

All manifestations of life . . . every state of being, every function, reaction, thought, experience, emotion or perception is a representation of the "on" or "off" phase of the programming of life on earth.

The Living Psyche

All life is psychic.

Life's psyche is its protector. It protects by overseeing and initiating all actions needed to preserve life.

Life in all of its forms has the universal problem of sustaining itself.

It must constantly supply itself with the food substances and oxygen needed to remain alive. It must use these to create its own protoplasm in order to repair itself, grow and reproduce.

The continuing struggle to obtain the essentials needed to sustain growth and maintenance is the universal life force directing all living organisms.

The Living Psyche of Plants

Plants are the unequivocal masters at providing for their needs. Their psyche provides them with a unique awareness of and affinity with the sun. They are also acutely aware of the availability of water and carbon dioxide in the atmosphere and the essential minerals they need in the soil.

Plants are psychically aware of everything they need to carry out photosynthesis and to create their own food and oxygen.

Plants are the most psychically aware of the On-Off presence and power of the sun. They are the earth's perfect species living in absolute On-Off harmony with the sun.

The Living Psyche of Animals

The psychic awareness of animals is no longer concerned with sunlight and the need o carry out photosynthesis for survival.

The animal has lost its affinity with the sun and its ability to perform photosynthesis. It is unable to create its own food and source of energy and must "steal them" from other living organisms in order to live.

Psyche must make the animal aware of the availability of its prey. It must also warn the animal of the presence and danger of its predator.

The psyche of plants, concerned with creating life, has been replaced by the Predator Instinct—Eat Or Be Eaten.

The Living Psyche of Man

The human psyche has an enormous task.

Unlike animals, the programming of the predator instinct in humans is incomplete and unorganized.

The human psyche must create ways to compensate for an instinctive program that is incapable of feeding and protecting us.

In humans the original predator program directing the hunting of prey in an environment adapted to both predator and prey is interrupted at birth and remains incomplete. The programming of our human predator instinct is cut short at birth and left woefully underdeveloped

We are born too soon. Unlike all other living creatures, we are born too early. We are born when our brain and physiological systems are only beginning to develop.

Our "prey" and territory are not yet defined. We lack the complete predatory species programming of other animals—and like animals we lack the species programming perfection epitomized by plants.

We are fortunate, however, to have inherited a superior brain endowed with enormously creative psychic power. Our ability to solve survival problems by thinking evolved as compensation for our lack of an efficient predatory program.

The expanded, creative ability to think was critical for survival and fortunately the human brain had evolved to a highly developed state.

The early birth that had interrupted the organization of the predator instinct had also virtually eliminated the original territory suitable for survival.

We believe human thought evolved from this need to create ways to feed ourselves and to avoid being captured and eaten. This also had to be accomplished in a frightening new world to which we had no programmed adaptation.

Human thought developed to compensate for our inability to feed and protect ourselves. Human thought needed to create ways to compensate for our defective predator instinct—ways to eat and prevent being eaten.*

---

* Reference: *Are We Humans Born Too Soon?*

The Beginnings of Psychic Awareness

We have established that a psychic awareness of the self is the defining characteristic of all life. Without this awareness there is no life. We humans say that our soul has departed.

Can this psychic awareness of the self be equated with life's original awareness of the presence of the sunlight and the phenomenon of photosynthesis?

Can we therefore conclude that psychic perception, the force that characterizes and is unique to all life, began with the awakening of the plant in the sunlight and became expressed in the process of photosynthesis?

————

Forms of Psychic Awareness

The psychic awareness unique to all life appears to take two forms. First and probably the more fundamental, appears to be the awareness of the sun exhibited by plants. Plants possess an extraordinary degree of psychic awareness of the sun.

The second form of psychic awareness is the individual organism's sense of itself—and its ability to sense things that are safe and beneficial or dangerous and harmful to its self.

Every form of life, whether plant or animal, microscopic or multi-cellular, possesses this individual sense of self. Each "knows" instinctively how to sustain itself and how to protect itself from being destroyed.

Any prevention to this instinct for self-preservation will cause the organism's eventual death. As Freud established, the repression of this sense of self in humans, for example, will cause psychic disorders and mental illness.

This psychic sense of "what's good for the self" is unique to life and differentiates the living from the non-living.

Psychic awareness appears to have begun when a single-celled, plant-like organism became aware of the power of sunlight.

This tiny form contained chlorophyll and was able to bond chemically with the sun's energy and apply it to a process known as photosynthesis.

It was this mysterious interaction with the sun that "turned on" the force of life.

The On-Off Theory of Life maintains that it is the power of the sun that governs all of the physiological and psychic functions necessary to life's continued existence.

We suggest that life's psychic sense of self originates in its sense of the sun.

We believe the degree of the completeness of a species' programming for survival reflects the degree of its relationship with the sun.

Plants compose the species that is most completely programmed for survival and consequently for immortality. They live most perfectly in harmony with the sun.

Animals are incompletely species programmed for survival. They are unable to use the power of the sun to create the food materials and energy needed to sustain themselves. They must exist by devouring other living organisms.

The human species is the species that is most incompletely programmed for survival. We are born too soon. We are born in a virtually helpless state with our brain and physiological systems incompletely developed. Years of physical and psychic care during childhood are needed for our survival. Unlike all other creature, the human development of the brain and physiological systems develop after birth. The unique too-early birth is essential because of the small size of the birth canal and other evolutionary factors. (To learn more about this extraordinary event refer to: *Are We Humans Born Too Soon?* by the authors.

Psychic Awareness in Plants and Animals

As previously noted, psychic awareness is the unique characteristic present in all living organisms. Its function is to support the organism's survival in every way possible. The nature of the awareness the organism possesses appears to be adapted to serve the organism's needs.

Following is a comparison of the psychic awareness of plants, animals and man.

Plant Awareness

Plants possess a completely efficient awareness for living. They are above all psychically aware of the presence of sunlight. They are aware of the availability of water, carbon dioxide in the atmosphere and the various minerals in the soil that they need.

The psyche of plants is also very creative. It can create a variety of adaptive devices such as thorns, poisonous excretions, etc. to protect itself from invading parasites and predators of all kinds.

The psychic awareness that directs photosynthesis and plant life is the most vital, natural and efficient system of life on earth.

Animal Awareness

The species programming of psychic awareness in animals changed radically when they lost their chlorophyll and became predators forced to hunt and eat other living organisms. Their psychic awareness changed drastically to accommodate this new more dangerous life style. They needed to become acutely psychically aware of the availability of prey and the presence of their predator.

In addition, with the change of environment from sea to land, the programming of survival of higher forms of animals required even more adaptations in their psychic awareness.

The species programming of the predator animal that was able to survive had to include a special awareness that allowed it to identify both its prey and its predator. In addition, all three—the animal, its prey and its predator—had to be species programmed for adaptation to live together in the same territorial environment.

Human Awareness

Born too soon, we humans have the most ineffective species programming of all life forms.

We are born to exist as predators but our programming is totally inadequate as a system for survival.

However, we fortunately possess a superior brain and a psyche that is incredibly creative. They provide us with the ability to create necessary illusions about ourselves, which are needed during our "make believe" childhood development. These exercises in play and role-model identification psychically organize us and eliminate the fearful state of helplessness resulting from our early birth.

To overcome our incomplete species programming, it is necessary that we create these extravagant illusions during childhood in fantasy and play. We must develop the psychic organization necessary to overcome the instinctual fear of the predator, the "bogey man".

During childhood, we psychically program ourselves to believe the illusions about our strength and ability. These fantasies will serve us well. They will provide the foundation for a strong sense of self and the ability to act aggressively on our own behalf.

Human psychic awareness is not completely species programmed as it is in animals. It must therefore be re-programmed during childhood using fantasy and illusion.

We must note, however, that the human program is still that of a predator and must serve the predator instinct. Effective psychic awareness must signal good conditions for capturing the prey and must also signal the danger of being captured by a predator.

Human survival as a predator required a very superior brain with the extraordinary creative ability together with a greatly expanded psychic awareness needed to solve problems.

Nature's Dedication To The Self

In nature each living organism exists solely and exclusively to serve itself.

Its species programming for survival excludes the needs of any other living organism—except for the nurturing of offspring. We believe that perhaps, during the short period after birth, the offspring may still psychically remain a part of the self.

Nature's dedication to the self is total and absolute.

Nature makes no compromises. Survival of the fittest rules. This is the natural way.

In the living world psyche leaves nothing to chance. The programming of the species that will survive must satisfy the organism's needs or the organism will cease to exist.

The Plant Psyche

In the world of plants, the photosynthetic process functions psychically in perfect harmony with the rotation of the earth and the on-off availability of sunlight.

The plant is able to psychically recognize and act aggressively with the availability of sunlight, water, atmospheric carbon dioxide and certain minerals in the soil.

The Animal Psyche

In the animal world photosynthesis has been replaced by the destructive predatory system.

In the animal's species programming for survival, psyche, in its role as protector, changed.

Psychic awareness of the animal now needed to clearly identify and instantly signal the presence of both prey and predator.

Successful programming also required that both the prey and predator live in, and be adapted to, the same environmental territory.

The Human Psyche

Note: Human species programming and the human manifestations of the predator instinct are discussed in detail in Chapter VI of this work.

How Psyche Protects

As previously noted all forms of life appear to be psychic. It is this ability that allows the organism to evaluate and initiate acts dedicated to survival and the preservation of its species.

The Psyche of Plants

The psyche of plants most importantly must locate sunlight. It must also find a supply of water, carbon dioxide and the essential sulfates and nitrates in the soil. The plant's psyche also protects by signaling unfavorable conditions such as the absence of sunlight, the shortage of other materials needed to carry on photosynthesis. It also needs to warn of the presence of destructive insects and other parasites.

The Psyche of Animals

The psyche of each animal species must learn how to identify its Prey and differentiate it from other animals and plants. The animal psyche must also be expert in the ability to detect and warn of the danger and presence of its predator.

The Human Psyche

The human psyche unfortunately has to overcome additional and even greater survival problems than those of any other living creature.

The human psyche faces the daunting task of repairing an incomplete species program. The genetic blueprint for acquiring food was no longer the simple act of hunting and capturing prey in a suitable territory.

Since human programming is incomplete at birth, our human infants are born in a mindless and helpless state. They are born before their brain is developed. There is awareness of only pain and pleasure. "Prey" and "territory" awareness are non-existent. Compared to all other predators, we humans are the most inadequately species programmed for survival.

―――――――――

The Development of Human Thought

The inefficiencies in our species programming left us helpless to even feed ourselves. It required us to compensate for the deficiencies if we were to survive. Solving the problem of helplessness required unique resourcefulness. It required the development of creative thought. Fortunately we humans had inherited a superior brain and a very creative psyche and so were able to overcome the obstacles to survive.

We believe that human thought developed to serve the predator instinct. Thinking was necessary to compensate for our inability to operate as a predator. We required a successful way to hunt, feed and protect ourselves.

———————

The Life-force—Aggression

Aggression is the universal force unique to all life. It is exclusive to each individual organism and directs the activities of the organism.

The aggressive life-force is entirely and absolutely dedicated to the preservation of the organism's individual self and to no other.

Even when it appears that an individual's aggression is directed toward another self's interest, it is actually satisfying and serving itself.

The aggressive instinct belongs exclusively to the individual's self and serves only that individual organism.

Prohibiting the free expression of this aggressive instinct in any living organism will severely damage or destroy it.

Prohibiting this aggressive instinct in the human child, for example, will result in repression of the self and mental illness.

———————

The Human Self

As previously stated, nature is absolutely dedicated to the preservation of the individual self and to no other. It applies to all life forms—plant and animal. Darwin noted the many ways it is expressed in the survival of each living species.

This law of nature also applies to us humans. If we are to be psychically healthy, we must act in our own interest as nature instinctively dictates. It is also true that when we are organized and healthy, we are capable of psychically understanding and supporting the well being of others. However, even these acts are probably experienced as psychically self-satisfying.

A great amount of support is needed by the human child to develop an independent and healthy sense of self. On the other hand, the recurring repression of this natural instinct will result in a lifetime of conflict and mental illness.

For a further exploration of this important unconditional law of nature and its influences on the development of the human personality, refer to Chapter VI—The Extraordinary Importance of Human Species Incompleteness.

CHAPTER V

The Elegant Creative Beauty of Plant Life

THE UNNATURAL ABERRANT LIFE OF ANIMALS

The Simple Beauty and Perfection of Plant Life

Plants are the most perfectly species programmed for life on earth. We believe them to be the "perfect" form of life.

As previously described, t hey function in perfect harmony with the earth's rotation and the on-off availability of the sun.

They are the creators of life using only what their environment freely offers—the energy of the sun, minerals in the soil, water and carbon dioxide in the atmosphere.

They possess a natural awareness of their environment and exist in a perfect relationship with the earth and the sun.

Plants are completely species-programmed to exist as long as the earth and the sun exist. They are theoretically immortal.

Authors' note: If one believes in God as the Creator of life, it is probable that He would be most proud of the creation of plant life.

————

The Unnatural Aberrant and Destructive
Life of Animals

Animals live unnaturally. They must destroy life to live. They possess a perverted awareness of their environment and have lost their original affinity and relationship with the earth and the sun.

The animal has become a destructive predator who requires the following imperfect and unnatural conditions to live:

It must exist in a specific and supportive territory to which it, along with a community of other animals and plants, exist. An ecological balance of predators and prey must be present.

It must have developed the psychic awareness to perceive and identify its prey and its predator. It must also be psychically aware of all other favorable or unfavorable survival conditions.

It must have developed the complex physcological systems needed to accommodate this abnormal new way of life. It must have evolved physiological systems able to ingest and assimilate an other living organism. It must be able to

capture it
engulf it
digest it
circulate and absorb its nutrients
convert these nutrients into protoplasm

> eliminate toxic waste
> repair itself, grow and reproduce.

In addition, it must have developed an effective psychic system of awareness to protect itself from being captured and consumed by another predator.

Authors' note: The life-style of the predator has no positive features. When the animal lost photosynthesis, the natural gift of the sun and the earth, it lost the essence of life. It lost the ability to create life. It became a destroyer.

Animal Survival

Animals who survive are completely species programmed as successful predators. Those that are not do not survive. To survive, the animal must be able to successfully capture and consume another living organism. The animal must be equally able to avoid being captured and consumed as the prey of another predator.

The animal kingdom (and man) exists exclusively as predator. All activity, except sexual activity, are dedicated to the fulfillment of the predator instinct—capturing and consuming another living organism and avoiding being captured and consumed.

The Unnatural Destructive and Incomplete
Life of Humans

We humans also live unnaturally. We too must destroy life in order to live.

We too have lost our original affinity and relationship with the earth and the sun and have become a destructive predator.

As a predator, our programming for survival is sadly incomplete since our species programming was interrupted by our abnormally early birth. We are born with predator instincts but with no programmed ability to carry them out.

We are born helpless. We have no idea of who we are. Our psychic awareness is incomplete and unorganized. We are unable to feed and protect ourselves. We are terrified of imaginary "ghost predators". We have no specified territory and no territorial adaptation.

How then did we survive our species deficiencies?

We survived by using our ability to think and to imagine.

Independent thought which exists outside of the species programming can only be possible when a very intelligent creature is born too soon causing an opening in the program.

Two things are therefore required for independent thought to develop. The creature must possess a highly developed brain capable of advanced function—and must be born too soon which canceled the remainder of the species program and allowed the brain to function independently. Psychic reactions such as emotions and advanced learning then become possible.

The too soon birth event was also necessary for early man to survive physically. Evolutionary changes had produced hominid ancestors with increasingly large heads and brains. These ancestors had also become erect, standing on two legs which allowed them a vastly expanded perspective of the world.

Unfortunately, however, the small birth canal had not changed. The newly erect body, the small birth canal and the larger head presented a serious birthing problem. For the mother and infant to survive, the infant has to pass though the canal and be born (too soon) before its head and brain could develop to its full size.

Our Human Predator Instinct

The predator instinct is based on the terror of being captured and eaten. As previously noted, our human predator instinct is incompletely programmed for survival.

Since we are born before our brain and physiological systems can be developed, the instinct for self-preservation is unorganized at birth. Our infants are born into a state of terrifying instinctual helplessness incapable of functioning successfully as a predator.

Unlike animals that have all inherited a completely organized instinctual program for feeding and protecting themselves, human infants are born with an incomplete program.

Fortunately however, we humans have inherited an unusually intelligent brain and highly creative psyche, which if given care and allowed to develop during our childhood years have the power to help us survive.

Children who are loved and cared for during their infancy and childhood will instinctively compensate for the deficiencies. They will replace the missing programming with imagination and create illusions regarding their strength and ability. There will be daunting heroes and gods, brave warriors and beautiful princesses with whom to identify.

These fantasy and play creations serve to overcome the infantile fear of helplessness. They provide the child with illusions regarding his own strength and ability needed to defeat the bogeymen and other lurking predatory creatures.

Because our predator instinct is so badly defined, we are left with vague memories of a cat-like predator—which we recreate in the form of evil spirits and devil gods.

It is interesting to note that our devil creations with their devil's tails and horns appear to resemble the ears and tail of the big cat predators of our chimpanzee relatives.

We humans have created an incomplete but highly sophisticated version of a predator lifestyle in order to survive. Although it operates in a very complex structured society and civilization, it still has its one objective: namely, to feed the self and to prevent the self from being captured and eaten.

———

CHAPTER VI

The Nature of Death

The Nature of Death

Death is not supposed to occur in the "true" form of life—the life that exists as a result of the power of the sun and the process of photosynthesis.

This, the "true form of life" continuously cares for itself. It produces its own food from basic naturally occurring compounds in the earth and utilizes its unique photosynthetic relationship with the sun.

A continuing method of providing for life has been established. Only interference with the process can cause a cessation of life in "death." The programming of life does not include death.

Death, as we know it, is an anomaly. It is an unfortunate mistake that occurred with the appearance of the animal form of life.

This animal form of life is a radically predatory way to exist. Animals and man live entirely by consuming other forms of life. They create nothing—they simply consume and destroy.

Plants, on the other hand, are creators of life.

It appears that animals and man evolved by means of an enormously inefficient method of existence. The method seems destined to break down resulting in inevitable death. In contrast, plant life is beautifully efficient, exists in harmony with the forces of the universe and is destined to live indefinitely.

What Then is the Nature of Death?

Death appears to be the final result of an extremely inefficient method of existing by using very bad machinery. As previously noted, animal life seems to have evolved in a "topsy turvy" fashion, experimentally adding and eliminating "parts" as they contributed to survival.

The creation of this makeshift machinery was required to process and produce "food" from the captured bodies of other living organisms. In addition, it was necessary to provide the energy needed to fuel the machinery. Devises in the form of gills and lungs had to be designed to capture the oxygen produced and released by plant life.

Without doubt, the bodies of animals and man re the most astonishingly bizarre of nature's creations.

As previously noted, there were an extraordinary number of problems to solve in order to survive in the predatory world.

Without doubt, the bodies of animals and man are the most astonishingly bizarre of nature's creations.

As previously noted, there were an extraordinary number of problems to solve in order to survive in the predatory world.

There is the problem of capturing suitable prey without damage to oneself. This is followed by the necessity to engulf it properly without damage to oneself. Then it has to be successfully digested, again without the danger of digesting oneself. The digested nutritious matter has then to be separated and identified as food. Another system is needed to distribute it to all parts of the body.

A system for obtaining a continuous supply of oxygen has to be created. This system must also convert oxygen into the chemical energy needed to carry out these functions.

Still another system is necessary to eliminate toxic waste material.

Chemical processes had to be devised for the production of the innumerable chemical compounds known as enzymes, hormones and others (many still unknown) needed to carry out the complex functions.

Finally, each cell in the body had to be capable of converting the food substances into its own special genetic protoplasm.

It is interesting to note that to this day, their basic evolutionary processes exist and are evidenced in the cellular processes of all animals and man.

For example, the systems used by the single celled animal protozoan of both ancient and present times are remarkably similar to the systems of the cells of all other animals both invertebrate and vertebrate including man.

All of the various systems devised to accommodate the predatory animal existence are remarkably similar attesting to their common evolutionary history.

The Nature of Death

We cannot say that we understand the nature of death as it occurs in animals and man.

We can perhaps surmise, however, that the corrupt and self-destructive system of living might eventually break down and result in death. Perhaps the body through misuse and aging simply ceases to operate?

The Psychic Meaning of Original Sin

An imponderable question to ponder . . .

Does the negative, destructive predatory way of life carry a kind of psychic sense of its own abnormality?

After all, it does represent a fundamental "betrayal" and rejection of its true nature. And because of our incomplete programming are we therefore psychically aware that we deny our true nature?

Do we identify it as "original sin?"

Do we sense our incompleteness and predatory way of life, not just as a benign defect but also as an "evil" deviation from the natural "good?"

Do we feel that we are not just incomplete,

but are an abnormal and sinful anomaly?

Do we somehow sense the basic integrity of nature's creative way of life and therefore loathe our predatory deviation from it? Is this why we pray for our salvation and forgiveness of our sins?

CHAPTER VII

The Extraordinary Importance
of Human Incompleteness

The Extraordinary Importance Of The Human Psyche

The extraordinary importance of our species incompleteness is apparent in the enormous problems it creates. Consider the problems we face as a species . . .

We are predators. We are born with the predator instinct as our only means of survival—the instinct to Eat Or Be Eaten.

We are born before our predator instinct programming can become organized and functional.

We are born lacking the essentials needed to survive as successful predators.

We are born with predator instincts but are helpless to execute them—and we are terrified of our helplessness.

Here are the problems we must overcome:

> Born too soon in a helpless condition
> Born with no psychic sense of self
> Psychically and physically unable to feed ourselves
> Psychically and physically unable to protect ourselves
> With brain and physiological systems undeveloped
> With no specific territory in which to live
> With no adaptation to our environment
> Terrified at our helplessness

No Psychic Sense of Self

We are born with no psychic sense of who we are. We are born with no sense of self. We must learn that we are human by imitating other humans. (The famous case of the Wolf Child comes to mind).

Unable to Feed Ourselves

Since our psychic and physiological programming is incomplete, our prey is not psychically identified. Even if we were able to capture prey, we would be unable to eat and digest it.

Unable to Protect Ourselves

We are born with no idea who our predators are. We have only vague images of faceless terrifying ghosts, goblins and bogeymen. We are born with no instinctive program of defense against a predator.

Undeveloped Brain and Physiological Systems

Unique in all of nature, we are born with our brain and metabolic systems undeveloped. We are relatively helpless as they develop and become organized during the many years of childhood.

No Territory and No Predatory Adaptation

We are not born with an adaptation to a territorial environment. We are born into a frightening, unfamiliar world with no territorial boundaries.

How did we overcome these incredible obstacles? How did we humans survive?

We survived by using our psychic capacity to create imagination and fantasy. We created substitutes for the missing information in our species programming. These childhood make-believe images probably emanate from unconscious memories of our predator evolutionary history.

The Human Psyche

As previously noted the psychic "self" of all living organisms whether they be plant or animal is absolutely dedicated to the support and survival of that life form *and to no other.*

The human psyche is no different. It is the great protector of each individual self—dedicated to support the individual's instinctual need to survive under any condition.

Psyche will use any method available to prevent physical or mental breakdown. If an individual is subjected to prolonged severe mental abuse, his psyche will create illusions and even delusions to ensure his survival.

Neurosis and psychosis are pathological programs for survival.

The human psyche acts as a psychic meter. It registers the body's physiological and mental state in the form of highly developed emotional responses. The emotions we humans feel are protective devices. Pleasurable emotions signal good healthful conditions. Painful emotions signal conditions dangerous to our health and welfare.

Early Man's Survival

Early man managed to get a foothold in the ancient world and to survive.

A famous paleontologist once remarked that man survived because he didn't taste very good and although comparatively defenseless was a slow breeder, could come down from the shelter of the trees, learn to stand erect and develop an opposable thumb.

We believe he survived because he had developed a superior brain with an imaginative and creative psyche.

He was also the sexiest hominid around having lost his prominent frontal lobe allowing him seductive eye contact. His sexiness also became more alluring since he had also lost most of his body hair except that showcasing the genitals.

———

The Species Programming of Our Close Relatives

These were still completely species programmed as animals:

Homo habilis
Homo rudolfensis
Homo ergaster
Homo erectus
Homo hydelbergensis

These became incompletely species Programmed like humans:'

Homo neanderthalensis
Homo sapiens

How Did We Survive?

Following is a summary of how we humans survived by using our psychic powers of imagination and creativity.

Overcoming Our Helplessness

We humans created the skill of parenting to help the child learn how to overcome his deficiencies. Good parent models help the child overcome helplessness by using imitation and imagination.

Overcoming Our Lack of Sense of Self

The child must create his missing psychic sense of himself. Only his sense of self can serve his instinct for self-preservation.

The self he creates must also identify him as human. He can create this image of himself by identifying with other humans as role models.

He must also develop a more complex and permanent image of himself. He will do this by developing psychic and emotional opinions about himself which will depend upon how he has been treated during his early childhood. If he has been loved and approved of, he will love and approve of himself.

The child who has been loved and allowed to develop a strong sense of self will become an emotionally healthy and independent person capable of acting aggressively on his own behalf.

Learning To Protect Ourself

This problem is also solved by use of creative fantasy.

Since the predator is unknown to us, he must first be created in order to be destroyed.

This is exactly what the normally developing child does again and again in fantasy and play. He creates many make-believe monsters whom he "successfully defeats" again and again in fantasy. He does this by using fictional and real human models to imitate and guide him as he achieves his great successes.

This child will become a mentally healthy and happy adult. Others less fortunate will be destined to remain in the infantile helpless psychic state fearing the predator who remains undefeated in their unconscious.

Because of our incomplete programming, the predator of childhood cannot be completely destroyed. He lives in our unconscious and makes his appearance in many ways in the human adult world. His fearful influence and "evil" presence is revealed in many forms. He, the undefeated memory of a long-lost prehistoric predator, especially persists in the widespread belief in the independent existence of good and evil. He is the originator of the idea of original sin.

The predator is ubiquitous in our literature and arts. He dominates all of our religions and spiritual beliefs. Note that he often appears in devil form with horns and a tail reminiscent of the distant memories of the ears and tail of the big cat predators of our ancestors.

Creating Our Lost Territory

Since we are born without a precise territory in which to live, we were obliged to create it and then attempt to live in it.

This vital deficiency in our species program left us with a permanent psychic sense of incompleteness and a compulsion to correct it.

So we organized and created family units, neighborhoods and local organizations. We continued to search for completeness expanding into larger towns, cities and countries. We created governments, cultures and societies.

Finally we explore the universe in a never-ending search for the missing instinctual territory which we can never find.

―――――――

Does Species Completeness Temporarily Occur in Sexual Orgasm?

We believe that it is possible that we humans are actually able to organized ourselves and achieve a temporary state of species completeness.

We believe organization may be possible at one time—during reproduction when the reproductive force of life is experienced. We believe a temporary state of species completeness occurs and is experienced in the fulfilling ecstasy of sexual orgasm.

―――――――

Awareness of Our Incompleteness

Although we have the advantage of independent thought, it can never replace the instinctual power of our species programming. Instinctual programming always absolutely rules.

We humans psychically sense that we are incomplete. It is experienced as a dangerous weakness which we constantly attempt to correct by organizing ourselves.

We compensate by organizing all aspects of our lives.

When we are organized, we feel less incomplete psychically and can experience pleasure and emotional well-being. When our organization is threatened, our sense of incompleteness increases and we experience various degrees of fear and anger.

We sense our incompleteness in degrees of psychic awareness. When we are well organized, we feel very well. The opposite is true when we become disorganized. We will suffer anxiety, fear and even illness.

Our pleasure and painful emotional responses depend upon our physical and psychic organization. The more organized, the less danger of incompleteness, and the more pleasure we experience.

Our Infantile Fear of Helplessness

Born too soon, we humans are unique in all of nature.

We suffer from the instinctual fear of being unable to feed ourselves and defend ourselves from "our predator". Remember that we are instinctively programmed to live as predators. In order to survive, we must capture and eat other living organisms and we must avoid being captured and eaten by a predator.

The inability to use our aggression in capturing prey and the fear of being eaten are the source of our infantile terror of helplessness.

Animals live as predators but they are not born too soon and are completely species programmed for survival as predators. They are programmed to hunt for food and to defend themselves against capture by a predator. Through millions of years of natural selection, they also have the advantage of being adapted to their environment.

Human infants instinctively react to their helplessness with terror and aggressive frustration. Unlike animals, the human predator instinct is not adequately programmed nor organized. We humans require years of "psychic programming" during childhood using fantasy and play to overcome our fear of helplessness. Only with good parenting support, heroic role models and make-believe games as guidelines, can the child psychically defeat the predator "bogey man" and develop the strength of a mentally healthy adult.

Conversely, the child who will become neurotic or psychotic will have been denied this vital support during the critical childhood

period. The terror of the predator will remain and will unfortunately
be transferred to the parent on whom the child will become psychically
dependent.

Fear will also force this child to repress his natural sense of self.
The repression will produce powerful rage as the self denies itself
and "hides" in the unconscious unable to experience the pleasures of
freedom and fulfillment.

The psychic program of fear and dependency will unfortunately
be permanent. The child will become a mentally ill adult who will be
psychically dependent and driven by fear throughout his lifetime.

The Human Psyche's Use of Fantasy

It is important to note that our psychic ability was no longer
absolutely restricted to species programming. Independent thought
(free of the program) was now possible.

However, it was absolutely necessary that the program losses so
vital to survival as a predator be replaced. So the human psyche created
incredibly ingenious systems to satisfy these requirements. It created
very advanced methods of capturing and amassing great numbers of
living organisms for our consumption as "food".

It devised elaborate and complex systems needed to locate, destroy
and protect us from our predator enemies.

It compensated for the absence of a programmed territory by
creating the tools needed to live in the vast and terrifying world in
which we found ourselves.

And equally important, the human psyche was required to cope
with, and overcome the rage and terror of the psychic helplessness we
experienced as a result of our incomplete programming.

And so the human psyche freely and endlessly created imaginary "stories" to explain away our fear of being captured and consumed by terrifying predators in this strange new world.

In addition, since we now had the ability to think independently, we began to wonder about the world around us and began searching for answers to the great mysteries beyond our ability to comprehend.

We are all aware of the myriad of "beliefs" that mankind has created in the search for answers to these great mysteries. These various beliefs, many in the form of major religions, are an important part of human thought and exert a powerful influence on human action.

They no doubt serve the deeply compelling psychic yearning we have to find the "completeness" resulting from our defective and incomplete birth.

Although these religious beliefs are valuable and psychically needed, it should be noted that they are actually the natural creations of our human psyche.

They serve the predator instinct symbolically and offer comfort and assistance to us in overcoming our fear of helplessness.

The Extraordinary Importance of Fantasy

Factors to consider:

- Humans are born with incomplete species programming.
- Due to the incomplete programming, the human brain is not developed at birth and must develop after birth. This is unique to the human species. All other organisms are born, hatch or germinate with their brain or psychic system fully developed and ready to function.

- When it is fully developed, however, the human brain is remarkably advanced.
- Since human species programming is incomplete at birth, a "window" to reality—a glimpse of the real world becomes possible as the brain develops free of the original programming.
- This window to an unprogrammed world offers the brain "food for imagination."
- The absence of a precise species program makes it possible for the human brain to develop advanced psychic ability and the capacity to create images based on imagination and fantasy.
- Since human beings are born completely helpless, drastic accommodations must be made for the race to survive.
- This restoration of the program must satisfy all of the problems needed to survive in the predatory world. It must also be able to psychically overcome the great new terror of helplessness and fear of the predator.
- In order to survive, substitutions must be made for the missing elements in the program.
- But how can there be substitutions for species programming? It would appear that nature and the Creator had not provided for this unpredicted human dilemma.

So how do we humans manage to survive the helplessly inadequate state in which we are born?

The amazing answer is that through our "window" of incompleteness, we are able to use our psychic powers of imagination and fantasy to create missing elements vitally needed by our species program.

We are able to replace the missing elements in the program with substitutions created in our imagination.

Through fantasy and play during the years of our childhood we manage to overcome our fear of incompleteness and the threats of the "predator bogeyman." During these childhood years we use fantasy and play to identify with heroic figures in order to overcome our infantile fear of helplessness and create a strong, independent psychic sense of our self.

We humans are able to accomplish this because we have inherited a distinctly superior brain. Unfortunately the brain is incompletely developed at birth and must become organized and functional after birth.

Since this phenomenon is unique to man, it will perhaps explain the presence of the extra dimensional "window to the world" we possess.

So although we humans are born with absolutely no species programming for survival and are certainly not equipped to survive as predators, we have managed to create extraordinary substitutions for our deficiencies.

It is ironic that perhaps it is because of our species deficiencies, that we have developed our unique psychic ability to create imagination.

And so we as infants are each required to begin the monumental task of creating our "human world" and "human identity." It will take us each approximately 18 years to accomplish this.

Since we have not inherited a species program which will specifically identify and protest us from our predator, we must devote many years to overcoming the childhood terror of being captured by an incompletely defined ghostlike bogeyman predator.

The terror of the predator—the fear of being captured and eaten is instinctive and absolute. The instinct to hunt for prey and the instinct to protect oneself from being hunted in turn together with the reproductive instinct are the total forces governing human and animal life.

All activities are dedicated to serving these three instinctual forces. There are no others.

Overcoming Our Incompleteness

We have shown that the human child must create trillions of images in his fantasy during the process of creating an image of himself as a human being.

If he is eventually to create an identity, he must receive a great deal of supportive care—both physiological and emotional. We describe the process as "childhood"—a period of play and make believe during which he pretends, experiments, imitates and tests himself in various imaginary roles in an effort to create his identity and sense of self.

To accomplish this, he will have to relive his psychic fantasies many millions of times. He must create a psychic self who is strong and relatively fearless. This "self" must be able to act aggressively on his own behalf and must be able to successfully function as a predator.

He, along with all other humans, will have become a creation of his own fantasy.

The World We Create Must Serve Our Instincts

We humans survive by creating a world made possible by our unprogrammed psychic capacity for imagination and fantasy.

We create this world to replace our lost world. And since we are predators, it is required to be absolutely dedicated to serve our predator instincts.

This world we created from our great capacity for imagination solely exists to satisfy our predatory and sexual instincts. It has no other purpose.

All of the astounding achievements of our world are mere replacement for the missing elements in our original predator program.

Even after childhood we continue to endlessly create replacements for our origingal species program.

We fulfill our instinct to hunt and capture prey by creating vast farms where we capture, cultivate and slaughter all manner of living plants and animals for our use as food.

We protect ourselves from the elements and predators by creating communities safe guarded by laws, police forces and armies of soldiers. We punish and execute enemies who threaten our safety.

We compensate for our sexually incomplete programming with a great variety of sexual enhancements—many of which are simple reproductions of childhood fantasies. The emotions we feel represent the various degrees of sexual incompleteness—sexual orgasm being the unique moment when species completeness may be experienced.

Our world of imagination and fantasy is primarily consumed with managing our pervasive fear of helplessness. Fear of helplessness is the single factor challenging our survival. It begins at birth and requires years to overcome and create a functioning human being.

The sense of incompleteness can never be overcome. Using illusion and fantasy only provides temporary diversion from the problem.

We create imaginary explanations in the form of philosophies and religions. We seek satisfaction in artistic creations and scientific inventions.

We pursue "knowledge" searching ceaselessly for answers to the unanswerable.

As predators, we humans are absolutely deviled to the preservation of ourselves. Self-interest is absolute. Any other act is the result of neurotic fear abnormally programmed during childhood.

The healthy human is dedicated to serving himself. Any other act directed at serving another (acts of compassion and charity) is psychic gratification that also serve the self.

CHAPTER VIII

Neurosis Revisited

Special Care Needed for Survival

We are born in the state of dangerous disorganization and must devote years of childhood using fantasy to create our missing nature.

As previously stated, the human infant is born helpless and incomplete. His brain and physiological functions are undeveloped and must become organized and functional after birth. This is unique to humans. The offspring of no other living species is born with an incomplete program for survival.

The normal human infant's painful experiences of disorganization are kept to a minimum by loving and supportive care. He is supported as he struggles to become physically and psychically organized and becomes able to use his aggression on his own behalf.

The infant who is destined to become emotionally ill will never be able to use his aggression on his own behalf. His efforts to develop normally will have been prevented by negligence or abuse. He will be forced to repress his normal aggression and will become psychically unable to survive on his own.

This unfortunate child will not become an independent and emotionally healthy adult. He will live by accommodating the parent figure on whom he will be psychically dependent. He will live with anxiety, fear and rage during his entire lifetime.

———

The Human Child

Consider that we humans are predators who are born too soon.

We are born inadequately programmed to survive in the predator world. Our infants are born physically helpless and psychically terrified of a vague predator against whom we are defenseless.

Under normal circumstances, the child will receive the care and support he needs to become psychically organized into an adult. He will be aided in overcoming his helplessness and instinctual fear of the "bogey man." This child will be encouraged to identify with strong role models and to use his own psychic creations in fantasy and play to conquer his fears.

The unfortunate child who will become neurotic will not receive the care and support he needs. He will be unable to develop psychically and will remain in the infantile helpless state of fearing the predator.

This child will not be able to develop into an independent adult. To survive, he will be compelled to remain psychically imprisoned in a lifetime program of dependency on the parent.

Both children began life in an equally inadequate state. Both were born helpless and with brain and physiological systems undeveloped. One will develop into a strong, mentally healthy adult. The other will not.

The other child will be forced to survive by repressing his sense of self and becoming psychically dependent upon the parent figure. The instinctual fear of the predator will be transferred to the parent. He will be forced to deny his identity and repress his struggle for independence. He will live his life in an infantile abusive dependency relationship with the parent figure whom he both fears as his predator and depends upon as his savior.

The Normal Human . . . The Mentally Ill Human
What Makes Them Different?

How the Normal Human Develops

As previously noted, the human infant is born much too soon, an event made necessary by the small size of the birth canal and other evolutionary changes. He is the only living creature who is born before his species programming is completed. Since his brain and physiological systems are not yet organized and functional, the infant must be fed and cared for or he will die.

Due to this untimely birth, the human infant has lost a significant part of his instinctual species programming. Since his brain has not yet developed, he is born physically and psychically helpless. Since he is alive however, he is able to experience terror. He is terrified when he is hungry and he is even terrified when "threatened" by a loud noise or the prick of a pin.

How then can this helpless infant survive and develop into a mentally and physically healthy adult?

He must first be provided with his physical and psychic needs by an emotionally healthy parent or caretaker. "His Majesty The Baby"* will be the reigning figure in this caring household.

Remember that we humans are born species programmed to live as predators but are inadequately prepared to function as predators.

---

*   Sigmund Freud, "On Narcissism"

How does the child develop a positive sense of himself and learn to act aggressively on his own behalf? How does he overcome his overwhelming infantile terror of helplessness? Of what is he so terrified?

The child is terrified of the instinctual memory of the predator—a formless, ghost-like ancestral creature. He is terrified of being captured and devoured by this bogeyman predator.

We humans are unlike animals whose predator and prey are well defined and who co-exist together, programmed for survival.

Fortunately although we humans have no instinctual defense against the phantoms and demons, we do have the means to create a defense.

We are fortunate to have a superior brain with an incredible psychic ability to create illusion. This human child who has no idea about why he is so afraid, will be able to create illusions that will serve to reassure and comfort him.

He cannot escape from the predator as an animal can because he doesn't know who his predator is. The only way he can eliminate the danger is to create the predator and then kill him.

This is exactly what the normally developing child does again and again in fantasy during his childhood years. He creates many make-believe monsters in play and successfully defeats them all. He also uses real and fictional role models to imitate and guide him as he psychically develops a self and a strong personal identity.

This child will become a mentally healthy adult able to act independently on his own behalf. (He will also incidentally become a successfully functioning predator.)

How the Mentally Ill Human Develops

How then does this child become a mentally ill adult?

Unfortunately this child does not receive the support he needs to freely create positive illusions about himself. His negative images are in fact increased by neglectful treatment.

This child is not supported in his attempts to create a positive and confident image of himself. He will be discouraged in his attempts to build an independent self, able to act aggressively on his own behalf. He will not overcome his infantile helplessness and fear of the predator.

Please remember that his parent-caretaker is himself emotionally dependent. He has also been prohibited from developing an independent self.

This unfortunate parent also remains in the helpless psychic state of dependency on his own parent. Since this parent is himself dependent, the child becomes a serious burden whom he must endure.

Remember that both are also born species programmed as predators. They are both born with inadequate survival systems that can only function after years of supportive care and positive self-building fantasy.

How then do they both survive?

Both child and parent are forced to repress their natural instincts for self-fulfillment in order to live.

How then can they identify and destroy the predator, as normal humans are able to do in fantasy and play?

The answer is that each of his or her "predators" has already been created for them. The child's predator is alive and has been transformed into the person of the parent. The parent's predator still lives in his psychic programming in the person of *his* parent.

Each experiences terrible emotional conflict since neither one can "destroy" his predator without destroying his only way of living. Each must remain helplessly dependent to survive. Each must live with, and under the control of, his own equally dependent parent.

Psychic dependency on the parent is established during infancy and early childhood and will remain unchanged during the individual's entire lifetime. All of his life's activities will be modified to serve this dependency relationship.

He will never develop a sense of himself and never be able to act aggressively in his own interest. He will remain psychically fearful and dependent on the predator parent whom he will both "hate and love".

Both child and parent were never able to "psychically destroy" the predator in fantasy. The abnormal dependency relationship of each with his parent is their only means of living. Each will suffer from mental disorders and will never be able to function as an independent, healthy adult.

Both child and parent will live in his own infantile helpless state of fear of the predator. The parent will remain dependent and fearful of *his* instinctual predator-parent. The child will live in a dependency relationship with the parent as his instinctual predator. Both will be unable to act aggressively on their own behalf. Both will be tormented by fear and rage in self-repression.

This psychic programming of dependency happens at one time only. It happens only during infancy *when the infant is truly dependent.*

This abnormal programming of dependency with the predator-parent is at the basis of mental disorder.

Repression, The Cause of Mental Illness

In the unnatural state of repression, the individual is forced to hide or repress his natural instinct for self-preservation.

He dares not display any sense of self nor act in any way on his own behalf. He must hide from the fearful parent who is himself emotionally ill and cannot care for the child.

Any independent expression or action of the child will become an extreme burden on this parent. For both to survive, the child must live in a selfless dependency and devotion to the parent.

The powerful natural instinct for self-preservation and self-fulfillment must be repressed. Both child and parent will each be a captive in his own repressive dependency program. Each will be in terrifying conflict between the two opposing forces of self-preservation and repression. It is this conflict that will cause both to be emotionally ill.

How The Neurotic Structures His Life

A neurotic person has been programmed to experiences terror whenever he acts on his own behalf.

His psychic programming of dependency with the parent began during infancy and continued throughout childhood. The dependency relationship tolerates no aggressive action on the child's part. This is an absolute requirement.

The child must therefore create a day-to-day program of living that will both protect his dependency on the parent and insure that there be no independently aggressive action on his part.

As the child develops into adulthood, this structured life of self-denial will cause him increasing painful conflict and mental anguish. Terror of losing the parent however, will compel him to go to any length to protect the dependency.

If there is a threat to the structured way he lives, and the threat increases, his terror of being destroyed by the parent-predator will increase and become intolerable. He will do anything to stop the terror including murder and/or suicide.

CHAPTER IX

The On-Off Programming Theory As It Applies To The Development of Human Thought and The Nature of Time

The Nature of Human Thought

We believe that plant life is the original form of life on earth and that the plant's method of living is a process programmed to respond totally to the on-off availability of the sun.

We maintain that all prevailing forms of life extant on earth with few primitive exceptions are descendant from plant life.

We recognize that the human being is born too soon—born before his species programming can be completely developed.

We believe that human thought is an anomaly.

The existence of human thought is possible only because of the break in the programming of the species.

Its nature consists of a mixture of perceptions—some based on fantasy, some on reality.

What we are unable to understand, we must create as fantasy in order to protect us from our terror of the unknown.

Animals who are survivors are completely programmed to respond only to reality situations in their lives. They are psychically programmed to respond to food, danger of a predator and reproductive impulse.

What therefore is human though? We humans fear the incompleteness into which we are cast—born too soon and helpless.

How much of thought is genuine representations of reality? How much is fantasy creations of an active psyche? Remember that the psyche of all living organisms is dedicated to protect life at all cost (even to create delusion as in the case of psychosis).

As previously noted, a great amount of fantasy is created during childhood to help the child organize himself psychically and to overcome his infantile terror of helplessness.

The human psyche probably creates a great deal of protective fantasy during a human's lifetime.

But what is the nature of thought?

We believe human thought is an inherited version of the on-off program of photosynthesis repeated in the on-off functioning of our brain cells.

All human thought is the product of the programming of the on-off availability of the sun.

Authors' note: The latest compendium of human thought, the computer, is a tribute to photosynthesis. It, like every human invention, functions according to the on-off program of life. Two symbols, zero and the number one, represent the on and off conditions of life and the availability of the sun. These are all that are required to communicate the entire of the human experience.

———

Human Emotions

Human emotions are creations of the human psyche used to express various degrees of pleasure and pain.

Emotions in the pleasurable category represent positive conditions for life—the "on" state. Emotions in the painful category signal conditions endangering life—the "off" state.

Our human emotions faithfully follow the on-off programming of photosynthesis and life on earth.

These psychic emotional reactions are protective "gifts" we inherited from of our plant ancestors. They are aberrant representations of the plant's wondrous natural awareness of the on-off presence of the sun.

Pleasure and Pain

All of life's responses including human psychic reactions are inherited from our plant ancestors. They instinctively recognize and respond to the on-off presence and absence of sunlight.

We believe the sensations of pleasure and pain are positive and negative instinctual responses to the presence and absences of the sun.

These instinctual responses we experience as human emotions are inherited from our plant ancestral responses during photosynthesis.

Pleasure and pain are psychic responses we experience as assessments of good and bad conditions affecting our well-being. They are psychic reproductions of our inherited responses to the positive and negative presence of the sun.

Pleasurable emotions signal positive conditions in the "on" state of the sun's presence. Painful emotions register dangerous conditions in the "off" state of the sun's absence.

The Nature of Time

We measure time by the number of on-off cyclic appearances the sun makes as the earth rotates and revolves around it.

But what *is* time? What is the nature of time?

Time, we believe, is the availability of the sun. Time represents the human experience of the availability of the sun.

Let's remember that plant life is the original form of life on earth and that the plant's method of existence involves the process known as photosynthesis. Photosynthesis is totally programmed in an absolute dependency with the on-off availability of the sun.

Plant life functions in a never-ending program based on the continued on-off appearances of the sun. Plants therefore exist in a never-ending program with time. Plant life is species programmed to be immortal.

When animals and man gave up photosynthesis and this harmonious unity with the sun and the earth, they lost their ability to live in a never-ending program with time.

Animals and man lost the on-off programming with the sun and time. They became destined to exist as predators and to eventually die.

Other Theories Regarding the Nature of Time

Authors' note: We believe that unless one starts with an objective non-human photosynthetic premise to explain the nature of time, one will be automatically transported into the human subjective world of psychic fantasy.

We believe the popular theories about the nature of time are in the human realm of psychic fantasy. Mc Taggert's Argument, Fatalism, Topology, Reductionism, Presentism, Platonism and Time Travel are excellent examples.

———————

CHAPTER X

Who Are We?

Why must we glorify ourselves and even deify our existence?

Why can't we accept ourselves as we are?

We humans are the only form of animal life that is aware of itself. Perhaps on some level of awareness, we sense that we represent a deviation from the natural form of life.

Perhaps we unconsciously perceive our predatory way of life as abnormal and repellant.

We don't seem to be able to accept ourselves as we are.

The fear and helplessness created by our incomplete species program appear to be too powerful to overcome.

Perhaps it is in an effort to compensate for this sense of inferiority, that we create illusions of superiority among us based on racial, religious, sexual and other differences.

Unfortunately, our relentless fear of helplessness will no doubt continue to keep us emotionally and everlastingly dependent upon our fantasies and delusions.

Authors' note: Let us remember the significance of our inherited photosynthetic origin. We are governed by the on-off availability of the sun.

When we are well, we are in the "on" state of harmony with the sun and life.

When we are ill, we have been programmed to live in an unnatural "off" death-like stage of existence.

Human Pleasure

We believe that the genuine sensation of human pleasure can only be experienced when the sense of self is free of the fear of the predator and is able to act aggressively on its own behalf.

If it is repressed and still hiding from the predator, a false form of pleasure is experienced as relief from fear of the predator.

In Conclusion

Let us remember the significance of our photosynthetic origin. We along with all life on earth function in on-off harmony with the earth's rotation and the on-off presence of the sun.

Extraordinary importance should be given to photosynthesis, the vital and mysterious process that enables green plants to begin the manufacture of their food and thereby become self-sustaining.

Plants may be considered the perfect form of life on earth. They are completely programmed for survival. They are immortal.

Animals and man, o the other hand, are an aberrant form of life, imperfectly programmed for survival. When animals evolved from plants, they lost their chlorophyll and their ability to produce their own food. Animals and man are predators reduced to prey on other forms of life to exist.

The existence of life on earth is totally rul ed by the on-off cycle of the sun.

The "on presence" of sunlight is instinctively understood by living organisms to represent the time to function and fulfill life's needs. The "off condition" signifies the cessation of functioning and potential danger of death.

Life can exist and function creatively only when the sun and the requirements for photosynthesis are attainable.

www.ingramcontent.com/pod-product-compliance
Lightning Source LLC
Chambersburg PA
CBHW031236280526
45784CB00004B/1597